Keratinocyte Biology - Structure and Function in the Epidermis

Edited by Mayumi Komine

Published in London, United Kingdom

IntechOpen

Supporting open minds since 2005

Keratinocyte Biology - Structure and Function in the Epidermis
http://dx.doi.org/10.5772/intechopen.95189
Edited by Mayumi Komine

Contributors
Emi Sato, Shinichi Imafuku, Carol A. Heckman, Razib MD. Hossain, Mayumi Komine, Miho Sashikawa-Kimura, Tuba M. Ansary, Koji Kamiya, Mamitaro Ohtsuki, Hidetoshi Tsuda, Jin Meijuan, Shin-ichi Tominaga

Notice
Statements and opinions expressed in the chapters are these of the individual contributors and not necessarily those of the editors or publisher. No responsibility is accepted for the accuracy of information contained in the published chapters. The publisher assumes no responsibility for any damage or injury to persons or property arising out of the use of any materials, instructions, methods or ideas contained in the book.

First published in London, United Kingdom, 2022 by IntechOpen
IntechOpen is the global imprint of INTECHOPEN LIMITED, registered in England and Wales, registration number: 11086078, 5 Princes Gate Court, London, SW7 2QJ, United Kingdom
Printed in Croatia

British Library Cataloguing-in-Publication Data
A catalogue record for this book is available from the British Library

Additional hard and PDF copies can be obtained from orders@intechopen.com

Keratinocyte Biology - Structure and Function in the Epidermis
Edited by Mayumi Komine
p. cm.
Print ISBN 978-1-80355-099-2
Online ISBN 978-1-80355-100-5
eBook (PDF) ISBN 978-1-80355-101-2

We are IntechOpen,
the world's leading publisher of
Open Access books
Built by scientists, for scientists

6,000+
Open access books available

146,000+
International authors and editors

185M+
Downloads

Our authors are among the

156
Countries delivered to

Top 1%
most cited scientists

12.2%
Contributors from top 500 universities

CLARIVATE ANALYTICS
BOOK
CITATION
INDEX
INDEXED

WEB OF SCIENCE™

Selection of our books indexed in the Book Citation Index (BKCI)
in Web of Science Core Collection™

Interested in publishing with us?
Contact book.department@intechopen.com

Numbers displayed above are based on latest data collected.
For more information visit www.intechopen.com

Meet the editor

Dr. Mayumi Komine obtained a Ph.D. and MD from the University of Tokyo in 1998. She investigated the molecular biological mechanism of keratin gene expression regulation at the Department of Dermatology, New York University, USA. Since 2018, she has been a professor in the Departments of Dermatology and Biochemistry at Jichi Medical University, Japan, where she is also the department chief and subject manager of the Department of Dermatology.

Contents

Preface

Keratinocytes are the main components of the epidermis, the outermost tissue of the human body. Epidermal keratinocytes are essential for maintaining the physiological function of the epidermis, which protects the inner body from environmental insults, including microorganisms, ultraviolet irradiation, mechanical injury, and thermal and chemical stresses. The biology of keratinocytes has been studied for a long time, and structural, chemical, and molecular studies have revealed their characteristic features. Keratinocytes respond to environmental stimuli by producing various cytokines, chemokines, antimicrobial peptides, and other molecules, inducing weak inflammation to fight against environmental stimuli; however, this inflammatory reaction ceases to maintain a normal physiological condition. Studies imply that keratinocytes have dual properties, as proinflammatory at the moment of insults, and anti-inflammatory at the later stage of inflammation. In some pathological conditions such as psoriasis and atopic dermatitis, subtle insults induce pathological pathways to develop new skin lesions because of the enhanced proinflammatory properties of keratinocytes. The anti-inflammatory property in these inflammatory conditions may be suppressed or defective and thus may result in enhanced inflammation.

It is unique in keratinocytes, and potentially other tissue cells such as endothelial cells and fibroblasts, to show both pro-inflammatory and anti-inflammatory properties because many immune cells are divided into subtypes that exhibit only certain aspects of immune functions, such as effector T cells that show proinflammatory function and regulatory T cells that show anti-inflammatory function.

It is important to study the anti-inflammatory aspects of keratinocytes and the mechanisms of ceasing the inflammatory reaction provoked by environmental stimuli to elucidate the mechanisms of maintaining the homeostasis of physiological conditions.

Section 1, composed of two chapters, discusses the structural components of keratinocytes, such as keratins and attachment molecules, and their importance in maintaining epidermal homeostasis. This section also reviews the importance of neighboring cells, such as melanocytes. Section 2 is composed of three chapters. The first chapter discusses the inflammatory aspects of keratinocytes as the main components of cutaneous phenotype formation in inflammatory skin diseases such as psoriasis and atopic dermatitis. The second chapter discusses stem cells in the epidermis and skin regeneration and aging as aspects of epidermal keratinocytes. The final chapter discusses keratinocytes and skin disorders, focusing on genetic abnormalities and barrier function of the epidermis and skin disorders.

It was an honor to edit this volume and I hope it will be a useful resource for those interested in keratinocyte biology and cutaneous disorders.

Mayumi Komine
Department of Dermatology,
Jichi Medical University,
Shimotsuke, Tochigi, Japan

Epidermis and its Structural Component

Cytokeratins of Tumorigenic and Highly Malignant Respiratory Tract Epithelial Cells

Carol A. Heckman

Abstract

In malignant airway epithelial cells, structural abnormalities were evident from the cytokeratin organization. To determine whether the cytokeratins themselves were responsible, an *in vitro* model for bronchogenic carcinoma, consisting of three highly malignant lines and three less tumorigenic lines, was studied. Cytokeratins were evaluated by two-dimensional polyacrylamide gel electrophoresis (2D-PAGE). When typical constraints on tumors were relieved by *in vitro* culture, lines showed profiles resembling normal, primary cells. The CK5/CK14 combination, characteristic of basal epithelial layers, was represented by CK6A/CK14. CK17 was invariably present, while CK5, CK7, CK8, CK19, and CK42 content varied. CK19 appeared to substitute for the rarely observed CK18. While lacking the common CK8/CK18 combination of hyperproliferative cells, an invasive, metastasizing line had CK6A/CK7 or CK8 with CK19 suggesting derivation similar to adenocarcinomas. Bands of CK19 and actin migrated to higher pI in tumorigenic and malignant lines than in normal cells. Ubiquitinated acidic cytokeratins with a low isoelectric point (pI) and high molecular weight (MW) showed no consistent differences in lines that differed in growth potential. Type II made up 49–52% of total cytokeratins in nonmalignant lines, whereas highly malignant lines showed lower levels. Posttranslational modifications were identified but could not explain the shortfall of basic cytokeratins.

Keywords: actin, motility, invasion, squamous cell carcinoma, metastasis, cytoskeleton, differentiation

1. Introduction

Intermediate filaments, which are made up of cytokeratins, are responsible for the structural integration and resiliency of the epithelial linings. To build up the 10-nm filament from the molecular level, a subunit is formed by the dimerization of one type I keratin and one type II molecule. These heterodimers attach in an antiparallel arrangement to compose the larger tetrameric subunit common to 10-nm filaments. End-to-end and side-to-side assembly gives rise to the long, flexible filaments seen in images of epithelial cells. When the large number of keratin genes is considered, plus their posttranslational modifications, there is an impressive variety of filament compositions. Each epithelial cell type is characterized by a combination of type I and type II cytokeratins. For example, some of the human cytokeratins discovered recently are highly expressed in the hair follicle [1]. While the cytokeratin profile of a cell depends on selective expression, which in turn

3

depends on its differentiated state, there is considerable latitude in expression profiles of some cell types.

It has long been suspected that the cytokeratins play a role in growth regulation. Indeed, CK8, CK17, and CK18 have been investigated with respect to their regulatory roles. Several basic or neutral type II proteins, including CK4–CK6, have a conserved site corresponding to Ser/Thr73 of the CK8 sequence. Posttranslational modification (PTM) at CK8 Ser/Thr73 was found downstream of proapoptotic receptor Fas/CD95-mediated c-Jun N-terminal kinase activation (see for review [2]). In different systems, phosphorylation of type II cytokeratins resulted in increased solubility of the filaments and/or collapse of the filament network. These are among the mechanisms contributing to dissolution of intermediate filaments during mitosis, which allows the cell to round up for division [3–5]. Knockout of the most commonly expressed pairs, CK5 and CK14 or CK8 and CK18, is often embryonic or neonatal lethal (see for review, [2]), affirming the importance of these keratins for epithelial cell differentiations in the lining tissues. Cell behavior was directly affected by cytokeratin content, as has been demonstrated by Magin and coworkers. Keratinocytes deficient in all the cytokeratins showed increased softness and invasiveness, which were largely restored by re-expression of CK5/CK14 [6].

The cytokeratins are also subject to complex transcriptional regulation. In the epidermis, the CK6/CK16 pair are induced within a day after injury [7, 8]. The stress-responsive CK6, CK16, and CK17 were all expressed in response to injury in skin and during hyperproliferation in psoriasis, suggesting their upregulation by the transcription factor, Nrf-2 (see for review [9, 10]). CK17 is regulated by several transcription factors as well as by the possible interaction of the ubiquitylated form with STAT3 (see for review [11]). High expression of CK17 in lung adenocarcinoma was predictive of poor overall survival, suggesting a close relationship to malignancy [12]. Evidence also suggested a process regulating CK6 through activator protein-1 binding, which may be regulated by c-fos and c-jun [13].

Type I also have a role in regulating cell growth. CK17 presence in the nucleus was shown to allow a complex to be formed by interaction with an integral membrane protein, LAP2β. This was thought to affect gene expression and cell proliferation [14]. CK18 was modified by phosphorylation at Ser 33 and Ser 52 sites, enabling it to interact with pathways regulated by parkin, a tumor suppressor [15]. Phosphorylated CK18 and CK19 interacted with 14-3-3 proteins, which promoted the solubility of cytokeratins and their recruitment to membranes [16, 17]. Whereas these modifications may modulate some of the non-mechanical functions of the cytokeratins, it is not known whether they affect growth. One possible way in which they could affect it is by regulating the formation of a CK8-Akt complex. It has been proposed that Akt binds to CK8 in the CK8/CK18 protofilament. Failing this interaction, it is hypophosphoryated specifically at a residue essential for activation. However, studies in which both CK8/CK18 were knocked down showed that Akt phosphorylation and activation were enhanced [18].

Previous studies have not addressed the question of whether any changes were related to oncogenic transformation but independent of the expression patterns associated with the tumors' differentiated state. Some 85% of human tumors originate from epithelial cells. As cytokeratins comprise a large fraction of the cytoskeleton, they could have an important role in growth control. Nevertheless, the profiles of tumors remained roughly similar to those in the tissue of origin (see for review [19]). Even in the normal state, the epithelial differentiation could be perturbed by chemical or physical agents, initiating changes in cytokeratin expression. For example, in upper airway epithelium, vitamin A deficiency caused the normal pseudostratified epithelium to be replaced by a metaplastic squamous epithelium [20]. This was accompanied by the disappearance of certain cytokeratins and a marked increase in expression of one or more others [21]. Both *in vivo* [22] and *in vitro* [23, 24], CK18 expression was reduced while expression of CK13, a marker for cornified squamous epithelium, was increased. Changes in type II cytokeratins were also found, especially enhanced expression of CK4, typical of stratified epithelial tissues (see for review [22]). Another change following toxin exposure was a reduction in CK15 expression in submucosal glands [25]. Upon removal of the pathological stimulus, the squamous metaplasia was reversed along with the cytokeratin profile (see for review [26]).

The cytokeratins of airway tumors were dramatically different from the composition of the normal epithelium. In parallel to the differentiation of the epithelial lining into a squamous epithelium, tumors with a squamous differentiation expressed cytokeratins typical of squamous metaplasia, namely CK4 and CK10. In contrast, adenocarcinomas, which are thought to arise from hyperplastic adenomatous lesions, expressed keratins typical of mucous cells, namely CK7, CK8, and CK18 [27]. CK13 was present in most tumors showing squamous differentiation, while CK4 was found with CKs 7, 18, and 19 in adenocarcinomas of the lung and adenosquamous tumors [28]. These studies did not screen for "keratinocyte-type" K5/K14 pair which typically composed the filaments attached to hemidesmosomes and desmosomes [1], but these cytokeratins were also found in non-small cell lung cancer [29]. CK6 and, in some cases variable levels of CK14 and CK15, were identified in squamous cell carcinomas. Altogether, keratins CK4-CK8, 10, 13, 17, 18, and 19 were found in squamous cell carcinomas, although the exact pattern depended on the differentiated state of each tumor [30].

It was clear from the above studies that cytokeratin profiles from airway epithelium depended on differentiation to such an extent that it was difficult to infer a relationship to growth control. As the defects that enable epithelial cells to invade the submucosa and metastasize are of great interest, it was desirable to reinvestigate the cytokeratins' relationship to growth potential after having compensated for the cells' differentiated state. The cytokeratin profiles of epithelia originating from the same tissue source could be compared in a well-characterized *in vitro* model system for squamous cell carcinoma. In the tissue culture setting, physical barriers to expansion were removed, and cells entered the logarithmic phase of growth within 48 h of being subcultured. Moreover, the nutritional composition of the environment could be simplified by growing the cell lines in identical media. Three highly tumorigenic lines and three lines with lesser tumorigenic potential were used [31], and the question of whether cytokeratin profiles were related to the altered growth potential of epithelial cells was revisited.

2. Materials and methods

2.1 Primary cells and cell lines

Normal, primary epithelial cultures were grown out of tracheal explants from specific-pathogen-free, inbred F344 rats. The cultures were grown in a Waymouth's medium enriched with amino acids, putrescine, sodium pyruvate, hydrocortisone, insulin, fetal bovine serum, penicillin, and streptomycin [32, 33]. Cell lines were also from the upper airway epithelium of F344 rats. Nonmalignant lines were derived after treatment with 7,12-dimethylbenz(a)anthracene- or 12-0-tetradeca-noylphorbol-13-acetate. They were tested in immune-suppressed host animals. They were found to be nontumorigenic at early passages but became tumorigenic after prolonged growth *in vitro* [31, 34–36]. Malignant cell lines were generated by treating tracheal tissues with benzo[a]pyrene (B2-1 and BP3) or

3-methylcholanthrene (MCA7). The malignancy of the resulting tumors was increased further by serial passage in animals [37].

2.2 Characterization of malignancy and differentiation

To classify the cell lines by their growth potential, we determined the number of cells required to induce tumors in 50% of the animals tested (T.D.50). Cultured cells were injected into the thighs of immune-deficient rats or athymic nude mice, as previously described [31]. The highly malignant cell lines were tested by titrating the dose down from 1000 cells and determining the frequency with which tumors appeared by 43 weeks. These lines all produced invasive, keratinizing squamous cell carcinomas. To check for metastases, 1×10^4 cells were injected into the thigh, and the leg with the primary tumor was amputated 6 weeks after the injection date. Animals were euthanized and necropsied 6 weeks after the amputation.

Tissue differentiation was studied by removing tumors at 7–13 weeks after injection and preparing them for histological examination. They were bisected, fixed in 10% buffered formalin, and embedded in paraffin. Sections were cut at 6 µm thickness, mounted on slides and stained with hematoxylin and eosin.

The tendency of cells to undergo terminal keratinization *in vitro* was evaluated by examining cells dislodged from the surface of confluent cultures. A stream of medium was directed across the surface of the culture and the suspended cells were recovered by centrifugation at 1200 × g. Smears prepared from the pellets were fixed in 95% ethanol and stained by the Papanicolaou procedure [38].

2.3 Extraction of cytoskeletal proteins

To prepare the cytokeratins for 2D-PAGE, cells were plated at a density of 7–12 $\times 10^5$ per 100 mm dish and allowed to become confluent. A cytoskeletal preparation rich in keratins was made by the extraction procedure of Franke and coworkers [39]. Samples for isoelectric focusing were made by rinsing the dishes twice with TNM buffer (140 mM NaCl, 5 mM $MgCl_2$, 10 mM Tris-HCl, pH 7.6) and then treating them with 1% Triton X-100 for 4 min. The detergent-extracted cells were then washed twice with TMN, harvested with a rubber policeman, and pelleted by centrifugation at 1500 × g at 4°C. An additional extraction was performed in some preparations. The culture dishes were rinsed with high salt TNM buffer containing 1.5 M KCl and 0.5% Triton X-100 for 10 min, and then the preparation was completed as above. The pellets were resuspended in sample buffer (2% SDS, 10% glycerol, 5% β-mercaptoethanol in 25 mM Tris-HCl, pH 8.3) and boiled until solubilized. Each sample was dialysed against 0.1 mM phenylmethylsulfonylfluoride at 4°C. The dialysate was lyophilized and resolubilized in lysis buffer (9.S M urea, 2% Nonidet P-40, and 5% β-mercaptoethanol).

2.4 2D-PAGE and quantification of Coomassie blue staining intensity

Separation of proteins by their isoelectric point was performed by the method of O'Farrell [40]. Samples containing 300 µg of protein were run for 18–20 h in a gradient made up of LKB Ampholine pH 3.5–10. The pH profile in the first dimension was determined by measuring the pH in small sections of gels processed in parallel with those containing protein samples. After separation in the first dimension, each sample was electrophoresed into a 4% polyacrylamide stacking gel and 10% polyacrylamide resolving gel in a final concentration of 0.1% sodium dodecyl sulfate (SDS). For each cell line, 6–10 gels were run. In addition, some 150 µg samples were run to confirm the identity of the major protein species. The gels were fixed in 12% trichloroacetic acid and stained with Coomassie blue, as previously described [40]. A lane of markers was added for the electrophoretic separation, including phosphorylase B (92.5 kDa), bovine serum albumin (66.2), ovalbumin (45.0), carbonic anhydrase (31.0) and soybean trypsin inhibitor (20.1). To analyze the total mass of cytoskeletal protein, samples were prepared with high-salt extraction, which ensured greater contrast between proteins and the background. A representative digital image from each line was selected, and the integrated absorbance at each spot on the gel estimated by comparison to the markers. After background subtraction, the boundaries around the spots were drawn manually, and the mass of each protein estimated by converting absorbance into intensity values, as previously described [41]. Total mass varied from ~60 to ~200 µg per gel.

2.5 Liquid chromatography-mass spectrometry-mass-spectrometry

For analysis by LC-MS-MS, proteins were excised from the gels, destained, reduced, alkylated, and trypsin-digested by a standard in-gel method. The peptides from each sample were concentrated and desalted using C18 Zip-Tip and reconstituted in 0.1% formic acid. Peptide mixtures were loaded onto a peptide trap cartridge and eluted onto a reversed-phase PicoFrit column (New Objective, Woburn, MA) as described elsewhere [42]. The eluted peptides were ionized and sprayed into the mass spectrometer, using a Nanospray Flex Ion Source ES071 (Thermo) and analyzed using a Thermo Scientific Q-Exactive hybrid Quadrupole-Orbitrap Mass Spectrometer and a Thermo Dionex UltiMate 3000 RSLCnano System.

Raw data files were searched against the rat protein sequences database using Proteome Discoverer 1.4 software (Thermo, San Jose, CA) based on the SEQUEST algorithm. Carbamidomethylation (+57.021 Da) of cysteines was set as a fixed modification, and oxidation/+15.995 Da (M), deamidated/+0.984 Da (N, Q), acetyl/+42.011 Da (K), phospho/+79.966 Da (S, T, Y), and ubiquitin-K/+114.043 Da (K) were set as dynamic modifications. The minimum peptide length was specified as five amino acids, and precursor mass tolerance was set to 15 ppm. The peptides' sequences and counts of peptide spectrum matches (PSM) were assembled into a Proteome Discoverer Report [42]. The effect of posttranslational modifications (PTMs) on the pI and MW of each protein were modeled using the Prot pi online bioinformatics tool [43].

2.6 Frequency and area analysis on gel spots and bands

For frequency analysis, spots on each gel were traced, converted to a binary image, and their areas were analyzed using the Particle Analysis module of ImageJ [44]. Under the assumption that each gel spot was a thin, continuous protein layer, the area was used to represent fractional volumes of each protein on the gel.

3. Results

3.1 Growth and differentiation of nonmalignant and malignant lines

Lines from the respiratory airway epithelium were divided into two groups on the basis of their abnormal growth potential as defined by their T.D.50 values. Tumor production required a 1000-fold greater number of nonmalignant cells than malignant cells (**Tables 1** and **2**). Moreover, none of the nonmalignant lines showed metastasis. Upon histological examination, all of the tumors had characteristics of invasive squamous cell carcinomas. The entire range of variation was represented

Tumorigenicity tests on cell lines of low malignant potential			
Cell line (passage)	Number of animals tested	Number with tumors	
		5×10^5	2×10^6
4C9 (17)	3	0/2	0/1
4C9 (38)	4	2/2	1/2
165S (16)	4	0/2	0/2
165S (32)	4	0/2	1/2
2C1 (13)	3	0/1	0/2
2C1 (23)	3	0/1	0/2

Table 1.
Cells of low malignancy tested in athymic nude mice.

Tumorigenicity tests on cell lines of high malignant potential				
Cell line	Animals tested	Number with tumors		
		1×10^2	3×10^2	1×10^3
B2-1 (37)	25	5/10	8/10	5/5
BP3-0 (1)	15	5/5	5/5	5/5
MCA7 (21)	30	9/10	10/10	10/10

Table 2.
Cells of high malignancy tested in immune-suppressed, isogenic F44 rats.

by the B2-1 and BP3 lines, with MCA7 tumors showing intermediate histology between these two extremes. The keratinized cells of B2-1 tumors were squamous in shape but rarely if ever became enucleated. Even terminally differentiated cells showed small, pyknotic nuclei (**Figure 1a**). Internal portions of the B2-1 and MCA7 tumors became necrotic and also showed areas of infiltration by lymphatic cells. In BP3 tumors, multiple layers of enucleated squames were readily formed (**Figure 1b**). All three lines metastasized to the lymph nodes, and B2-1 also metastasized to the lung in 60% of the animals. The morphology of the metastases resembled that of the primary tumors (**Figure 1c**).

After injection with the specified number of cells, animals were maintained for 24–43 weeks. MCA7, the most immunogenic of the highly malignant cell lines, was tested similarly and formed tumors in 2 of 2 mice after injection of 1×10^3 cells. Modified from [31].

After injection with the specified number of cells, animals were maintained for up to 22 weeks. 165S (T15) was tested in 5 irradiated host animals at a dose of 1×10^3 cells and formed a tumor in one animal. The other nonmalignant lines were tested under similar conditions but failed to produce tumors. Modified from [31].

Normal and tumorigenic cells *in vitro* showed similar tendencies to form stratified squamous epithelia at high-density. The degree of squamous differentiation in confluent cultures could be assessed by imaging cells exfoliated from the cultures. In normal, primary cultures, the exfoliated cells were squamous in shape but did not become enucleated. The exfoliated cells from the lines differed in shape, but exfoliated cells from both primary cultures and cell lines commonly had pycnotic nuclei (**Figure 1D–G**).

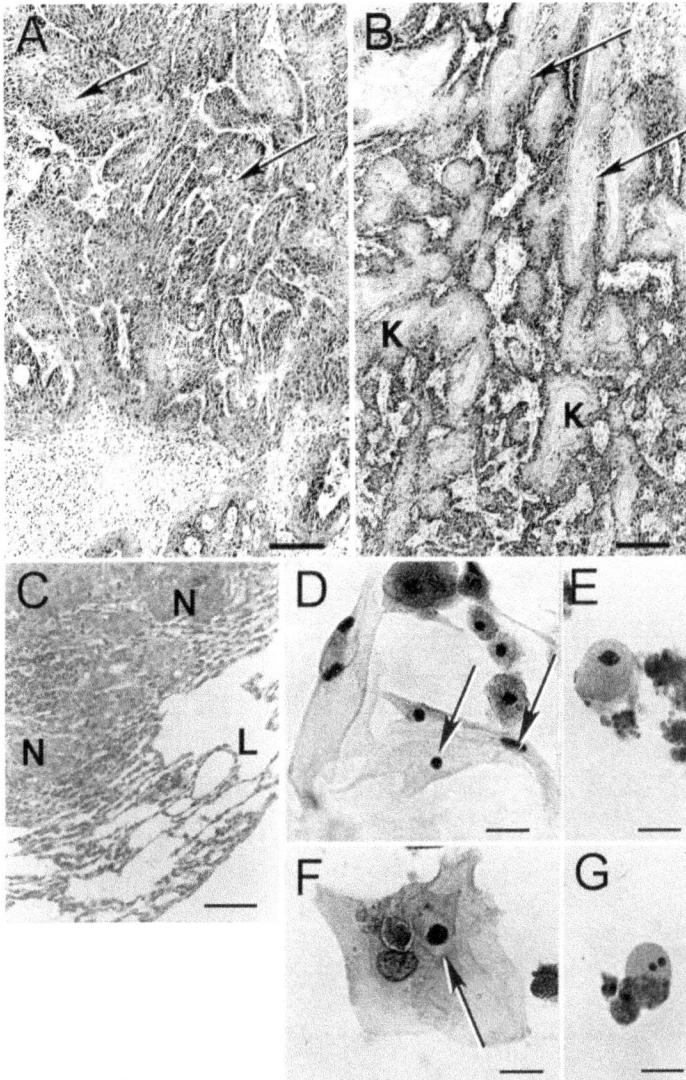

Figure 1.
Morphology of tumorigenic cells in vivo and in vitro. (A) B2-1 tumor. The squamous cells in the outermost epithelial layer exhibit pycnotic nuclei (arrows). Invasion into the surrounding tissue containing blood vessels (bottom) is obvious. (B) BP3 tumor. Squamous cells of the differentiated epithelial layers show enucleated cells (arrows) resembling keratin "pearls" (K). Cells are invading into the mesenchyme at the bottom. (C) Border of a B2-1 metastasis to the lung. Nests of tumor cells (N) form at the boundary with the compressed lung tissue (L). (D–G) Exfoliated cells. (D) Squames from normal cell cultures with pycnotic nuclei (arrows), (E) BP3 with pyknotic nuclei, (F) 165S cells with pyknotic nuclei (arrow), (G) 4C9 cell. Bars, (A) and (B) 500 μm, (C) 200 μm, (D–G) 20 μm.

3.2 Cytokeratins of normal epithelial cultures

Previous studies provided information about the keratin proteins of different cell types in the respiratory tract [1, 25, 28, 30, 45–53]. Methods for generating primary cultures were also well-known, and such cultures maintained the ability to repopulate normal epithelia after several weeks of growth *in vitro* [54]. Thus, the cytokeratins of the upper airways are summarized in **Tables 3** and **4**, and their presence or absence determined for cell lines differing in growth potential.

MW kDa, pI*	MW (rat)	pI (rat)	PTM (UniProt)		From LC-MS-MS	
			MW	pI	MW	pI
CK14[†,‡]			meth R, 1 phos		3 phos, 2 Ac	
50, 5.3	52.68	5.09	53.76	5.04	53.01	4.88
CK15[†,‡,¶]			7 phos			
50, 4.9	48.87	4.81	49.43	4.60		
CK19[‡,§,¶,‖]			9 phos, 5 meth R		2 Ac	
40–44 4.6–5.2	44.64	5.23	45.51	4.56	44.72	5.10
CK16[‡]						
46, 5.1	51.61	5.12				
CK17[‡,§]			2 phos		3 phos, 8 Ac	
48, 5.1	48.12	4.97	48.28	4.89	48.62	4.64
CK10			6 phos			
56.5, 5.3	56.50	5.11	56.90	4.90		
CK18[†,§,¶]			4 phos			
45, 5.1–5.7	47.76	5.19	48.08	4.99		
CK42						
	50.21	5.10				
β-Actin			Met oxidation		4 Ac	
41.7 5.2–5.3	41.74	5.32	41.77	5.32	41.91	5.03

*The MW (kDa) and pI on the left column are for human keratins [1]. Proteins of the same family are identified by colored typeface [55]. The effect of PTMs is modeled as described in Section 2. Abbreviations used are phos, phosphorylation; meth, methyl; Ac, acetylation.
†Submucosal glands.
‡Basal cells.
§Clara cells.
¶Ciliated cells.
‖Cuboidal bronchiolar cells.

Table 3.
Type I proteins and β-actin between 4.6 and 5.3 pI compared to the human map.

Due to the fact that cytokeratins are highly conserved throughout evolution, it was possible to gather preliminary data by overlaying a map of the cytokeratins [1] on the proteins separated by 2D-PAGE. As shown in **Table 3**, the posttranslational modifications of CK14 and CK15 had minor effects on their position in 2D-PAGE, so they were used to center the map (**Figure 2a**). The spot representing CK17 (241 PSM) was confirmed by LC–MS–MS but also contained CK19 (99 PSM) and CK42 (97 PSM). The levels of the latter two appeared high as estimated by PSM, but were much lower than CK17 when the MS1 area of unique sequence was analyzed. These proteins, CK19 (89 PSM), CK17 (56 PSM), and CK42 (34 PSM), also made up the band at 42–43 kDa. CK42 was not displayed on the map of human cytokeratins [1], because it is a rodent cytokeratin that was lost in primates [56].

The CK5/CK14 pair was characteristic of basal cells of the normal mouse respiratory tract [52]. However, CK5 often occurred in the absence of CK14 in the human airway epithelium [57], and the latter was considered a marker for metaplasia [58]. Because primary tracheal cultures were largely made up of basal cells [59], the presence of the CK5/CK14 pair was anticipated. The CK14 spot was prominent, and the 56-kDa band contained both CK5 (311 PSM) and CK6A

MW kDa, pI*	MW (rat)	pI (rat)	PTM UniProt (1) LC-MS-MS (2)		From LC-MS-MS	
			MW	pI	MW	pI
CK4						
59, 7.3	57.67	7.24				
CK5[†]			(1) 11 phos (2) Met oxidation		3 Ac	
56, 7.4	61.83	7.37	62.71	5.22	61.95	6.20
			61.91	7.37		
CK6A[†]			(2) Met oxidation		8 Ac	
56, 7.8	59.25	7.62	59.33	7.62	59.58	5.72
CK7[†,‡,§]			(1) 8 phos		2 Ac	
36.7–54, 4.6–6.0	50.71	5.70	51.35	5.29	50.79	5.51
CK8[‡,§]			(1) 22 phos (2) Ac		5 Met oxidation	
49.7–54, 5.4–6.1	54.02	5.85	55.78	4.65	54.10	5.85
			54.10	5.60		

*The MW and pI on the left column are for human keratins [1]. Proteins of the same family are identified by colored typeface [55]. Abbreviations used are phos, phosphorylation; Ac, acetylation.
[†] Basal cells.
[‡] Clara cells.
[§] Ciliated cells.

Table 4.
Type II cytokeratins of upper airways corresponding to pIs 4.6–7.8 of human.

(272 PSM), as shown in **Figure 2b**. The amount of CK13, a marker of cornified squamous differentiation, was negligible, but another squamous cell marker, CK10, was sometimes present (**Figure 2a**).

Cells of the glands and gland ducts were previously shown to express CK14, CK15, CK18, and CK19 [45], and CK18 was also found in columnar and mucous cells [27]. CK8/CK18/CK19 characterized all columnar cells in the upper airways *in vivo*, including gland cells [28] and Clara cells [53], but were absent from basal cells of the bronchi [28, 45]. Whereas CK18 was not found in primary rat tracheal cultures, CK19 was present (**Figure 2b**). The band of 42–46 kDa spots contained both CK19 and CK42 and included β-actin. Although the theoretical range of β-actin pIs did not extend further than 5.3 (**Table 3**), β-actin in this band extended up to pI 6.1. CK8, which was shown to form heterodimers with both CK18 and CK19, was also present in primary cells. In addition, small amounts of CK7 were present (**Figure 2b**).

3.3 Is the cytokeratin profile changed by sequential passage of a cell line?

As oncogenic transformation generally gives rise to genomic instability, it is possible that sequential passage of a line might alter the cytokeratin expression. Samples from the 4C9 line were collected in passages 12–14 and compared to samples collected at passages 33 and 34. Proteins in the MW range of 45 kDa extended into more basic pIs in 4C9 than in normal cells, but this pattern was unchanged with passage levels. Similarly, bands of very low pI (less than 3.9) at approximately 61 and 55 kDA, were unaltered (cf. **Figure 2a** and **Figure 3a**).

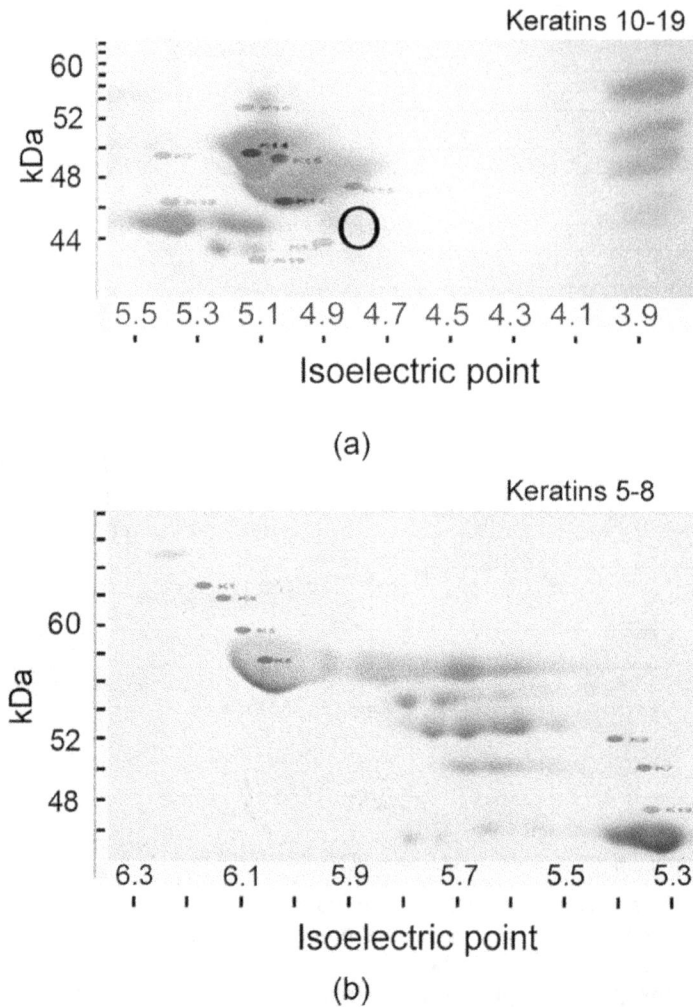

(a)

(b)

Figure 2.
Map of human cytokeratins [1] overlaid on a 2D-PAGE gel from normal tracheal epithelial cells. (a) Type I cytokeratins. There is a small spot at the site of CK10, and CK14 and CK17 are prominent. The proteins at 42– 46 kDa are β-actin and CK19 mixed with degradation products from CK8 and CK5. At a highly acidic pI, the ubiquitinated CK14 and CK17 proteins are present (see Table 5). (b) Type II cytokeratins. CK6A and CK5 are present at MW ~56 kDa, while low MW proteins, CK7 and CK8, migrate to lower pIs than other type II proteins. Proteins confirmed by LC-MS-MS analysis are underlined in Tables 3 and 4.

Although a 4C9 sample from passage 33 showed a trace of CK13, this was also present in samples from the early passages. K13 was expressed in squamous meta-plasia [22] and cultured tracheal cells [23]. The sporadic K13 spots observed in the current studies were consistent with the fact that, although the cells studied here showed signs of squamous differentiation, they were not cornified.

3.4 Differences in the actin and cytokeratins related to immortalization

The extension of the band containing actin and CK19 proteins into more basic pI ranges (cf. **Figures 2b** and **3a**) suggested that there may have been posttranslational modifications. Acetylation occurred on β-actin, as Lys61, Lys113, Lys213, Lys291, Lys315, Lys326, and Lys328 were found by LC-MS-MS. Modeling the addition of

Nonmalignant cell lines

(a)

(b)

Isoelectric point

(c)

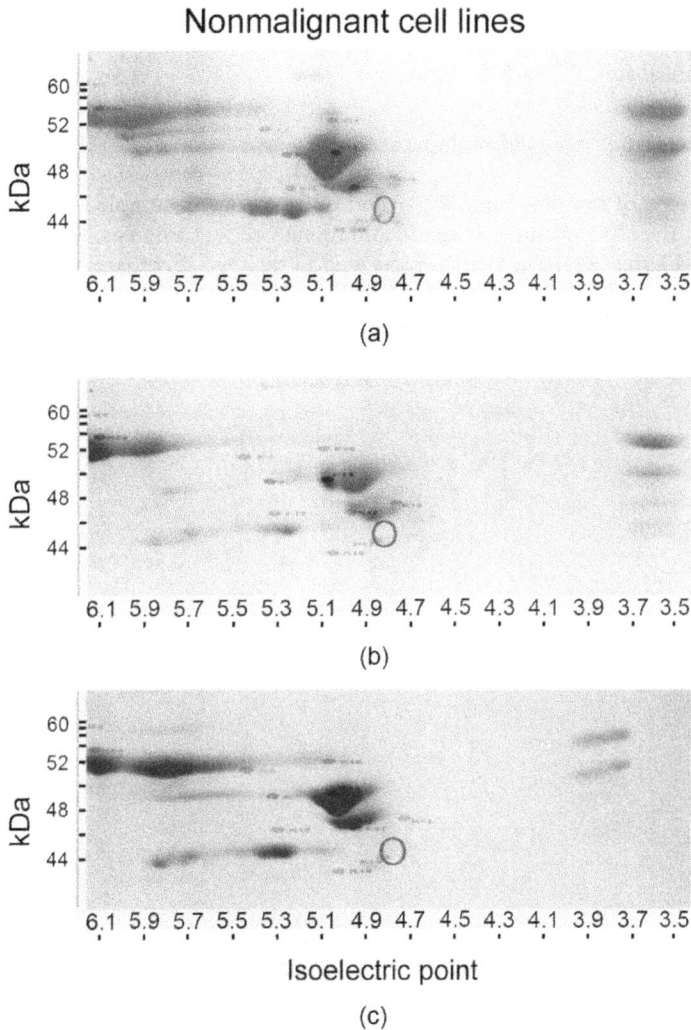

Figure 3.
*Cytokeratins from nonmalignant cell lines. The calculated MW is given in parentheses (see **Tables** 3 and 4). (a) 4C9. Type I cytokeratins, CK14 (53 kDa) and CK17 (48 kDa), are present along with actin and CK19 (below the marker for CK18) and type II CK5/CK6A (56 kDa) and CK8 (54 kDa). There is a faint spot in the region of CK13 (circled). Proteins present at high MW and pI of 3.5–3.9 are ubiquitinated CK14 and CK17. (b) 165S. The spot identified in (a) as CK14 (53 kDa) is resolved into two spots. CK7 (51 kDa) is present as a faint band below CK6A. (c) 2C1. The same cytokeratins are present as in (a) including a large amount of CK5/CK6A. As in (b), the β-actin/CK19 band extends beyond 5.8 pI.*

these residues together with Met sulfoxide modifications reported in the Uniprot database did not markedly change the predicted pI of β-actin, nor did any CK19 modifications found here shift the pI (**Table 3**). The 2C1 line, which had the lowest tumorigenic potential (**Table 1**), resembled 165S in the pIs of these 42–46 kDa proteins (cf. **Figure 3b** and **c**). The three lines showed slight differences in the cytokeratins of higher MW. In some 165S samples, the protein identified as CK14 was resolved into two spots (**Figure 3b**). The second was similar in MW but differed in pI, as would be expected for CK16. Likewise, in certain samples, the CK17 spot was resolved into two proteins. This is shown in a replicate 2D-PAGE experiment on the tumorigenic line (see Appendix, **Figure A1**). Trace amounts of

CK15 and CK16, found by LC-MS-MS analysis, were thought to reflect sporadic expression of these species. The spots sampled are shown in Appendix A (Portions of gel sampled for LC-MS-MS, **Figure A3**).

3.5 Cytokeratin profiles of malignant lines

The profile of samples from MCA7 and BP3 generally resembled normal cells. In particular, they had prominent bands containing CK8 (cf. **Figures 3b** and **4a-b**, and Appendix **Figure A2**). The mass represented in the type II cytokeratin, CK6A, declined in samples from all malignant lines, compared to primary and nonmalignant cells (cf. **Figures 3** and **4**). Unless the decrease in CK6A was

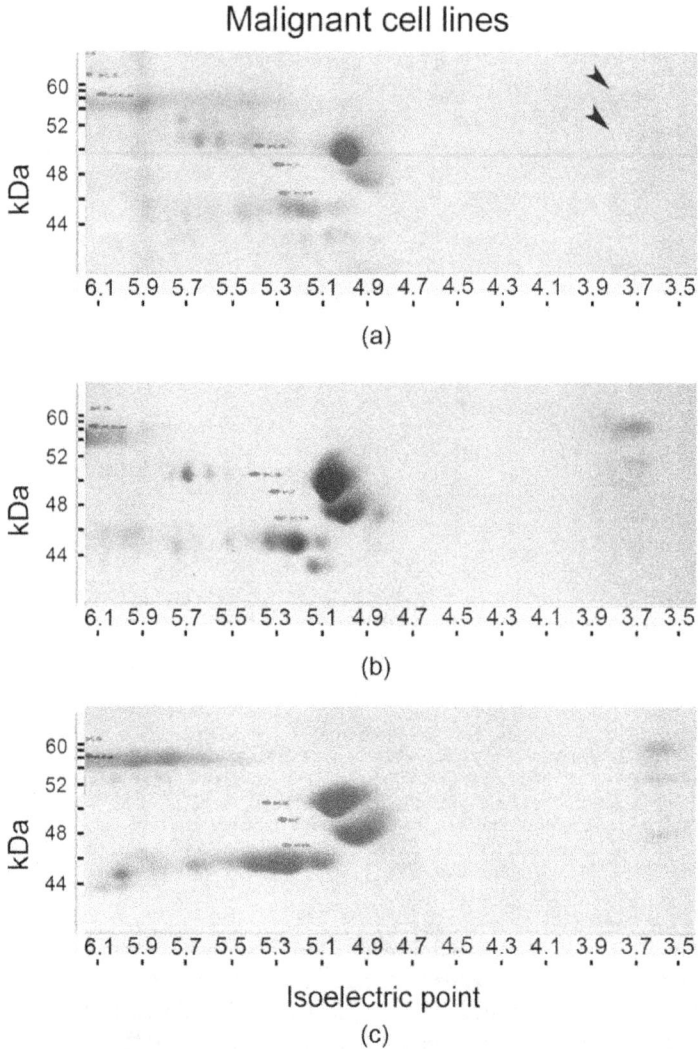

Malignant cell lines

(a)

(b)

Isoelectric point

(c)

Figure 4.
*Cytokeratins from malignant cell lines. The calculated MW is given in parentheses (**Tables 3** and **4**). (a) MCA7. The type I CK14, CK17, and K19 were mapped, along with type II cytokeratins, CK5/CK6A (62 kDa/59 kDa), and CK8 (54 kDa). The actin/CK19 band occupies a similar range of pI as in normal cells. Faint bands are visible at pI 3.7 (arrowheads). (b) BP3. The profile resembles that in (a), except that the actin/CK19 range extends beyond 5.9. (c) B2–1. The profile resembles (a) and (b), but CK8 is absent and proteins of the actin/CK19 band are increased.*

accompanied by an increase in another type II protein or a reduction in type I proteins, it would mean an imbalance between acidic and basic cytokeratins. This implication was investigated by determining the frequency of observing each major cytokeratin in the nonmalignant and malignant lines. The results suggested that the decline in CK6A content occurred over the entire range of pIs (uppermost markers, **Figure 5a**). There were also changes in a lower MW range within the pI range, 5.8–6.0. The actin/CK19 band was slightly more apparent in malignant cell samples at the extreme end of the pI range (lowermost markers, **Figure 5a**). With respect to the type I cytokeratins, two spots were occasionally observed at the location of CK17 in BP3 samples (**Figure 5b**). CK15 may have been resolved from CK17 at this location, as mentioned above. Results of replicate experiments on the highly malignant lines are shown in Appendix **Figure A2**.

The implication that there was a reduced content of basic cytokeratins was further investigated by selecting a representative 2D-PAGE image for each line and measuring the intensities of all spots. The results suggested that CK6A was the predominant type II cytokeratin in all samples except those from B2-1 where it varied widely. Thus, the total levels of type II closely followed those of CK6A. In the B2-1 line, CK7/CK8, and CK19 were elevated (**Figure 5c**). Heterodimers formed

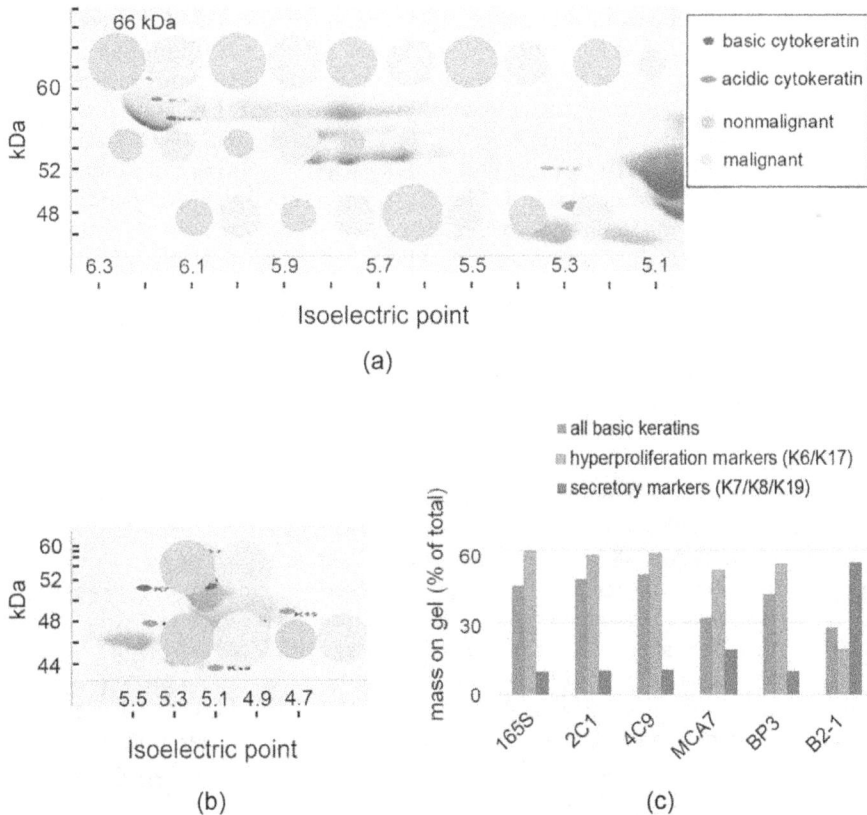

Figure 5.
Differences among the cytokeratins in selected samples from nonmalignant and malignant lines. (a, b) The frequency of observing a cytokeratin is represented by symbols for nonmalignant (green) and malignant (yellow). (a) Percentage of gels in which CK5/CK6A (uppermost), CK8 (middle), and the actin/CK19 band (lowermost) were present. (b) Percentage of gels in which CK14 (upper) and CK17 (lower) were present. (c) Content of type II cytokeratins and markers for secretory and hyperproliferating cells. Total type II content is similar to CK6A content except for the B2-1 line, where there is less CK6A. The results are representative of 27 gels.

between CK7/CK8 and CK19 were considered a marker for columnar or secretory cells (**Tables 3** and **4**). Interestingly, CK18 and CK19 were identified in sero (mucous) glands [28] and also found, along with CK7 and CK8, in large cell neuro-endocrine carcinomas *in vivo* [60]. This suggested that B2-1 arose from a differentiated mucous, neuroendocrine, or Clara cell. Nevertheless, the total mass of type I cytokeratins also exceeded type II in the MCA7 line, which was apparently not derived from a differentiated cell type. Thus, the shortfall in type II content was not restricted to the B2-1 line.

3.6 Is there a general difference between nonmalignant and malignant?

While the results of **Figure 5** suggested a decrease of CK6A, a larger mass of protein in the 42–46 kDa band would automatically have reduced the amount of CK6A apparent at certain pI intervals. This band made up 25% of the mass in 2D-PAGE preparations from malignant MCA7 cells. The mass of protein in high MW/low pI bands was also variable (cf. **Figures 2** and **3**). For images in which most spots were present, a quantitative analysis was performed. To this end, the integrated intensities of proteins in the gels were computed in ImageJ. The results showed that these ubiquitinated protein bands were present in all cell lines and made up 9–18% of the total. The comparison also made it clear that lower CK5/CK6A content in the B2-1 line (**Figure 5c**) was accompanied by greater CK8 content (**Figure 6a**).

It is widely thought that cells maintain a stoichiometry of 1:1 type I to type II, and if any imbalance occurs, that they are able to restore the balance (see for review [9, 61]). If the entire type II keratin gene cluster was eliminated, the CK14 expression level was also reduced [62], (see for review [10]), suggesting a dependency of type I on type II expression. As the laddered arrangement of high MW bands at low pIs (<3.9) suggested addition of 8.6 kDa subunits from ubiquitin, LC-MS-MS was performed on tryptic digests to determine whether these proteins showed SUMOylation or ubiquitination. The main constituents were CK14 (317 PSM), CK17 (179 PSM), CK42 (91 PSM), CK19 (76 PSM), CK12/13 (58 PSM), and CK15 (41 PSM). Some type II proteins, CK6a (67 PSM) and CK5 (47 PSM), were also found. SUMO1-3 were not found, but sites of ubiquitin addition on CK14 and CK17 were identified (**Table 5**). For CK14 Lys153 and CK17 Lys374, the same residue was alternatively acetylated or ubiquitinated. Most of the peptide represented was in one of these two forms, suggesting near-universal modification at those sites. One of the ubiquitinated sites in CK17 (K172) was homologous to that of CK19 (**Table 6**). Thus, although the bands at very low pI contained some basic cytokeratins, the shortfall of type II content could not be explained by selective ubiquitination, as these bands were mainly composed of type I proteins (**Table 5**).

The reason for the extremely low pI of these proteins was unclear. While both CK14 and CK17 are acetylated, modeling the effect of this PTM only slightly changed the estimated pI (**Table 3**). Modeling the addition of both acetyl groups and ubiquitin B onto CK17 gave a pI of ~5.1 and MW of ~74 kDa. The acidic pIs may have been due to PTMs not detected, or the proteins' configurations may have changed, causing basic charges to be sequestered in regions of the protein that were relatively resistant to unfolding.

In skin *in vivo*, ubiquitination of CK14 levels occurred in the basal cells, and so the above data were consistent with the known mechanism whereby Kelch protein, Cul3 substrate adaptor, recruited CK14 for ubiquitination [63]. As ubiquitinated cytokeratins were more soluble in Triton X-100 [64], the fractionation method of the current experiments could have resulted in artifactual loss of ubiquitinated proteins. As mentioned above for mass determinations, this could have accounted for a shortfall in type I, but not type II cytokeratins. Another possible source of

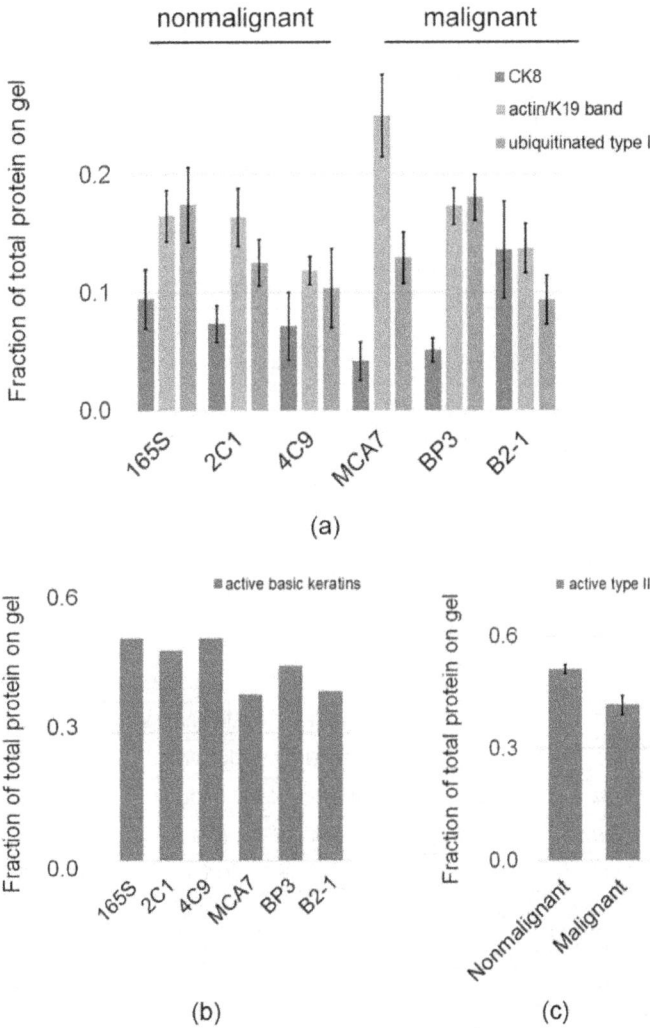

Figure 6.
Areas of spots identified on 2D-PAGE gels. The areas occupied were measured and represented as a fraction of the total area occupied by proteins on the gel. (a) Fraction occupied by CK8, bands of high MW and low pI, and proteins in the 42–46 kDa band. (b) Fraction of type II cytokeratins as a fraction of the total active cytokeratin, which was defined as the total area minus the area of the ubiquitinated cytokeratins. (c) Average integrated intensity of type II cytokeratins of nonmalignant versus malignant lines. Bars, ± standard error of the mean for 4–9 samples from each line.

artifact was phosphorylation, which again increased solubility and caused protein to be lost following dissolution of the filaments [4]. Although phosphosites were found in the current studies, they occurred in type I and type II alike (**Table 6**). As acetylation had a similar effect on solubility (see for review [10]), it is possible that the higher acetylation of CK6A accounted for some shortfall of the type II proteins.

Another source of artifact could be the variable mass of actin in the actin/CK19 band. Having been counted in the area of the acidic cytokeratins, the mass of actin could cause an artificial elevation in the mass counted as type I. MCA7 and 4C9 samples were at the opposite extremes of high and low 42–46 kDa content (**Figure 6a**). Despite being between these extreme values, BP3 cells still had less type II, suggesting that the actin band could not explain the effect overall. By assuming that the proteins at high MW and pI <3.9 were unavailable for filament

ID	Name	PTM	Peptide	With PTM (%)
Q6IFV1	CK14	Acetyl-K139	LATYLDKVR	54.5
		Acetyl-K153	ALEEANSDLEVKIR	83.6
		Ubi-K139	LATYLDKVR	4.0
		Ubi-K153	ALEEANSDLEVKIR	16.4
Q6IFU8	CK17	Acetyl-K172	TKFETEQALR	23.7
		Acetyl-K374	LLDVKTR	93.0
		Acetyl-K172	TKEFETEQALR	0.7%
		Ubi-K374	ILLDVKTR	2.5

*Ubiquitinated and acetylated sites of cytokeratins of high MW and low (<3.9) pI.

Table 5.
Ubiquitination in cytokeratin spots of high MW identified by LC-MS-MS.

ID	CK	PTM	(First residue)	Site
Q6IFV1	14	Phos	(7) QFTSSSSMKGSCGIGGGSSR	T9
	14	Phos	(209) TKFETEQSLRINVESDINGLR	T213, S216
Q6IFU8	17	Phos	(7) QFTSSSSIKGSSGLGGGSSR	S13
	17	Ac	(130) DYSAYYQTIEDLKNK	K142
	17	Ac	(202) ADLEMQIENLKEELAYLKK	K212, 219, 220
	17	Ac	(408) TIVEEVQDGKVISSR	K420
Q6IFU7	42	Phos	(316) SVQNLEIELQSQLSM	S316
Q63279	19/17	Ac	(94) LASYLDKVR	K100
Q63279	19/17	Ac	(170) TKFETEQALR	K171
Q63279	19/17	Ubi	(170) TKFETEQALR	K171
Q63279	19	Ac	(144) ILGATIENSK	K153
	19	Ac	(257) SQYEAMAEKNRK	K265
Q6P6Q2	5	Phos	(30) TTFSSVSR	T30
	5	Phos	(45) VSLGGAYGAGGYGSR	S46
	5	Ac	(60) SLYNVGGSKR	K68
	5	Ubi	(60) SLYNVGGSKR	K68
	5	Ac	(274) DVDAAYMNKVELEA	K282
	5	Ac	(426) NKLTELEEALQK	K427, 437
	5	Ac	(262) TTAENEFVMLKK	K272
Q6IG12	7	Ac	(324) AKLESSIAEAEEQGE	K326
Q6IG12	7/8	Ac	(182) TAAENEFVLLKK	K192
Q6IG12	7/8	Ac	(388) KLLEGEESR	K388
Q6IG12	7/8	Ac	(91) TLNNKFASFIDK	K102
Q10758	8	Ac	(317) LQAEIDALKGQR	K325
Q4FZU2	6A	Ac	(150) TEEREQIKTLNNK	K157
	6A	Ac	(172) FLEQQNKVLDTK	K178
	6A	Ac	(240) SKYEDEINRR	K241

ID	CK	PTM	(First residue)	Site
	6A	Ac	(250) TAAENEFVTLKK	K260, 261
	6A	Ac	(328) AQYEEIAKR	K335
	6A	Phos, Glyc	(339) AEAESWYQTKYEELQITAGR	S342, O-GalNAc
	6A	Ac	(365) NTKQEISEINR	K367
	6A	Ac	(380) LRSEIDHVKK	K388
	6A	Ac	(389) KQIANLQAAIAEAEQ	K389
	6A	Ac	(414) GKLEGLEDALQK	K415
	6A	Ac	(433) LLKEYQDLMNVK	K435
	6A	Ac	(523) GISSGLSSSGGSSSTIK	K539
	6A	Glyc	(523) GISSGLSSSGGSSSTIK	S525

*PTMs are on unique peptides unless otherwise designated. Ac = acetyl, Phos = phosphate, Glyc = O-GalNAc, Ubi = ubiquitin.

Table 6.
PTMs identified by LC-MS-MS in the cytokeratins of airway epithelium.

formation, it was possible to get an accurate estimate of the type I:type II ratio for each line. Type II proteins from nonmalignant cells made up 49–52% of the total (**Figure 6b** and **c**), suggesting a balance close to the hypothetical 1:1. In the malignant lines, basic cytokeratins constituted 39–46% of the total protein in 2D-PAGE gels. Despite its variability in malignant lines (**Figure 6c**), the lesser fractions suggested a shortfall of type II in the total cytokeratin fraction.

4. Discussion

Investigators' attempts to address the relationship between cytokeratins and growth control have been frustrated by the tremendous complexity of cytokeratin expression. In the current research, this was overcome by growing cells in media with the same composition and collecting them in the log phase of growth *in vitro*. This relieved the constraints of the *in vivo* environment. The cytokeratin profile became surprisingly similar to normal, primary cells, even for lines differing in growth potential *in vivo*. Despite the fact that the rat lines all formed squamous cell carcinomas upon testing in hosts, the cytokeratins typical of human airway squamous cell carcinomas *in vivo* [28] were much reduced. There was little representation of cornification markers, and CK4, the characteristic cytokeratin of stratified epithelia, was rare. In contrast, cytokeratins that were not prominent *in vivo*, e.g. CK6A and CK17, were present at high levels.

Thus, the conservatism of cytokeratin expression, which has made these proteins useful markers of a tumor's cell type of origin, was overcome by the *in vitro* conditions. Comparisons among the cultured lines suggested that B2-1 cells were derived from mucous cells or Clara cells, but otherwise there was a dearth of differentiation markers. For example, CK5/CK14 characterize the basal cells of compound epithelia, but the *in vitro* cultures showed CK5 levels that were far lower than CK14. The CK8/CK18 pair making up loose filaments, called "the simple-epithelial keratins", characterized the lumen linings of pseudostratified and complex epithelia (see for review [1]). This pair was found in most non-small cell lung cancer lines [65]. Whereas CK8 was present in most of the samples analyzed here, it was at lower levels than CK6A, except in the B2-1 line which apparently originated from a columnar cell type. In the airway epithelium *in vivo*, CK6 and CK16 were upregulated during development of squamous metaplasia [22, 66], as was CK14 in proliferating airway cells [52]. CK6A, typically paired with CK16, was a marker of hyperproliferation and wound healing, but the current results suggested that CK17 substituted for CK16 in the airway epithelial cultures. The CK6A/CK17 pair characterized both normal cells, nonmalignant, and malignant cell lines, suggesting that it was a hyperproliferation marker and unrelated to malignancy.

One novel observation from these studies was that ~10% or more of the cytokeratin mass is ubiquitinated. These species do not seem to have been reported previously but may have been missed because they occupy very low pI regions after 2D-PAGE. Although these bands might represent disordered proteins that are being recycled, only ubiquitin-mediated targeting for degradation has been reported to date [61, 63, 64]. As very little type II cytokeratin content was found in the highly acidic, ubiquitinated bands, turnover mainly affected type I proteins. A previous proposal, based on knockout of the type II genes, held that acidic cytokeratins were degraded rapidly in the absence of basic cytokeratins [67]. This is supported by the current work. The type I and II cytokeratin content is normally equal, but the contents here were not equally weighted. If the ubiquitinated protein content had been included in the mass estimates, an even higher type I:type II ratio would have been represented. The shortfall, however, became noticeable in the lines of high malignancy.

Previous evidence suggested that high levels of CK8 can facilitate tumor progression (see for review [9]). Although CK8 levels were elevated in some samples of the metastasizing cell line, B2-1, this may reflect its derivation from a differentiated secretory cell rather than oncogenic transformation. On the other hand, the species of 42–46 kDa and high pI, especially CK19, are of interest. The samples from tumorigenic and malignant lines generally differed in the migration of CK19 and/or actin to higher pI. PTMs leading to a high pI in CK19 were previously associated with worse prognoses in adenocarcinoma [51]. While subcellular reorganization of cytokeratins is typical of cells with invasive or malignant traits [6, 31], it remains unclear whether this is due to signaling or merely mechanical properties. Some of the cytokeratins were implicated in signaling, but clear relationships to growth control were lacking. CK18 and CK19 bound 14-3-3 proteins (see Introduction) and Src kinase was inhibited by its association with CK6 [68]. If type I cytokeratin expression depended on the presence of type II, as suggested above, it is possible that greater turnover of acidic proteins occurred. This may account for the current finding of ubiquitylated proteins at low pI. Thus, further studies on the deficit in 1:1 ratio of basic to acidic cytokeratins may shed light on the mechanism of structural revision during tumor progression.

Acknowledgements

The author is grateful for the technical assistance of Yatish Shah, Ronald Manger, Greg Ridder, Marcella Williams, Carrie Greenway, Jon Hao, and Jessica Barnett.

A. Appendix

A.1 Portions of gel sampled for LC-MS-MS

The identity of the proteins on gels was determined by reference to a 4C9 sample. Six positions on the gel, shown on **Figure A3**, were excised for LC-MS-MS. The MS raw files can be found at pCloud.

https://u.pcloud.link/publink/show?code=kZuGtUXZjqkXZG6pCyJ-TO3VJHoKAWna7dM8gSU9gV

4C9

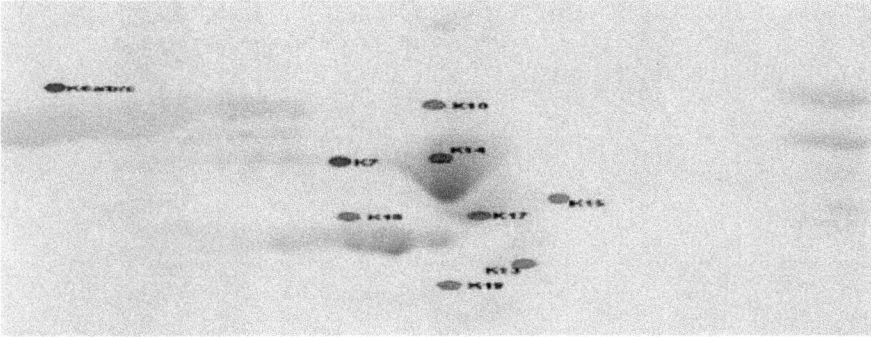

165S

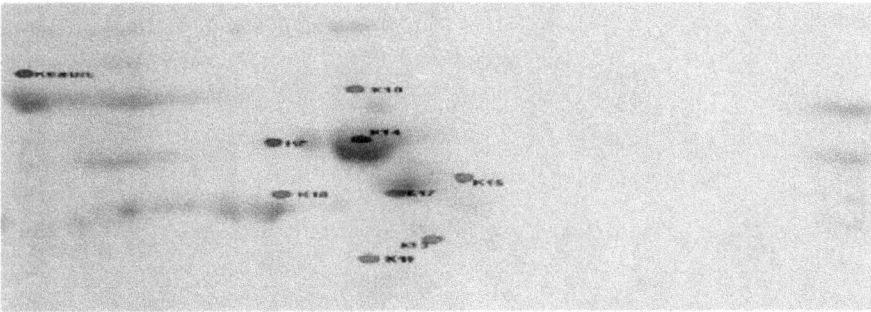

2C1

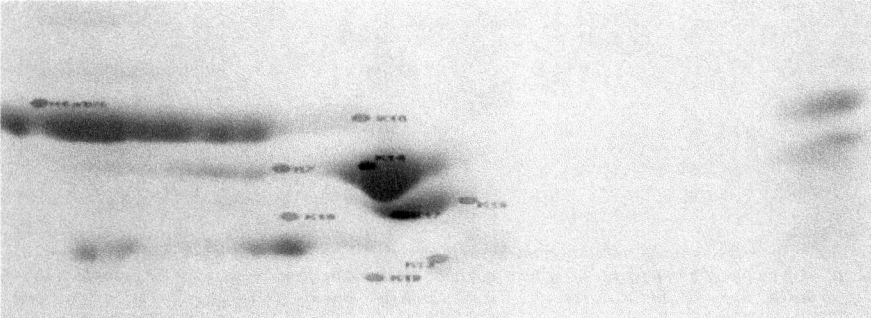

Figure A1.
Replicate experiments on cytokeratins from nonmalignant lines. In 4C9, CK14 (53 kDa) and CK17 (48 kDa) are present along with actin and CK19 (below the marker for human CK18) and type II CK5/CK6A (56 kDa) and CK8 (54 kDa). Bands are also present at high MW and low pI (3.5–3.9). In 165S, CK14 and CK17 are present, along with actin and CK19. Type II cytokeratins are represented by CK5/CK6A (56 kDa) and CK7 (51 kDa). In 2C1, the same cytokeratins are present as in 4C9.

MCA7

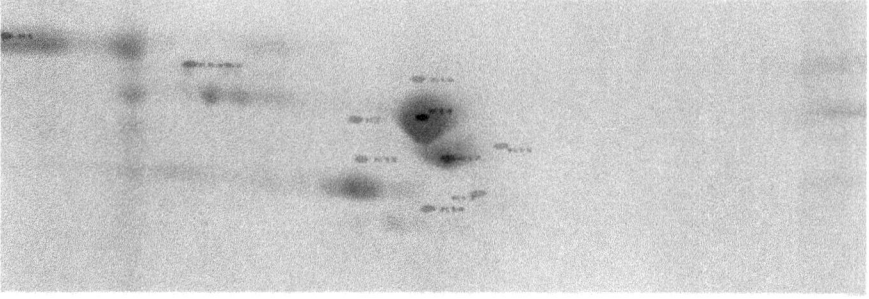

BP3

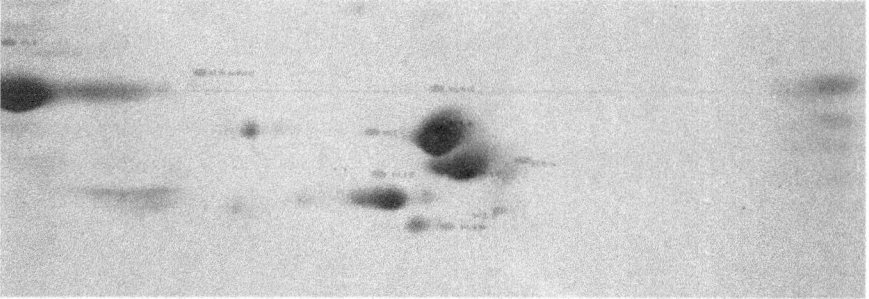

B2-1

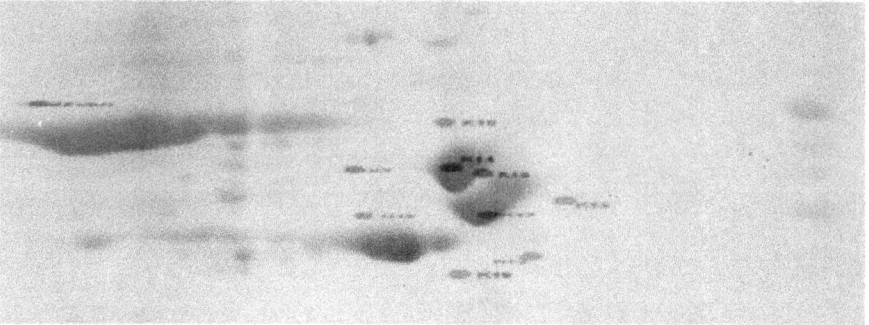

Figure A2.
Replicate experiments on cytokeratins from malignant lines. Type I cytokeratins, CK14 and CK17, were identified in all samples by reference to the map of human cytokeratins. Spots representing CK42 and CK19 are close to the CK18 marker. In MCA7, the type II cytokeratins found were CK5/CK6A (62 kDa/59 kDa), and CK8 (54 kDa). Faint bands are visible at low pI. In BP3, the profile is similar except that CK6A is the predominant type II cytokeratin. There is also a faint spot near the marker for CK15, as well as four bands at low pI. In B2-1, the CK8 band and proteins of the actin/CK42/CK19 band are prominent. Again, there are four discrete bands at low pI.

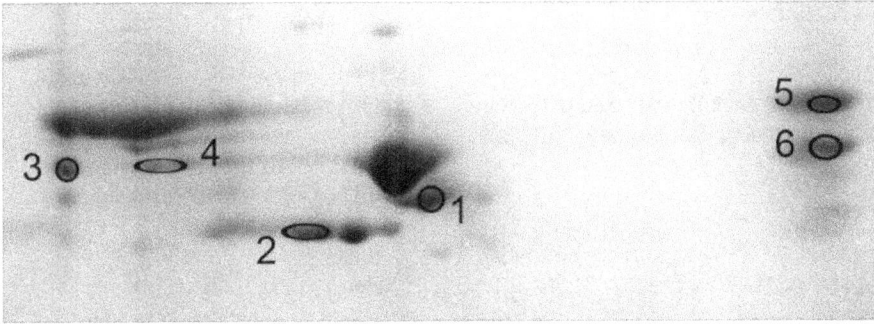

Figure A3.
Cytokeratins from 4C9 identified by LS-MS-MS.

Author details

Carol A. Heckman
Bowling Green State University, Bowling Green, Ohio, USA

*Address all correspondence to: heckman@bgsu.edu

IntechOpen

References

[1] Moll R, Divo M, Langbein L. The human keratins: Biology and pathology. Histochemistry Cell Biology. 2008;**129**: 705-733

[2] Karantza V. Keratins in health and cancer: More than mere epithelial cell markers. Oncogene. 2011;**30**(2): 127-138

[3] Toivola DM, Zhou Q, English LS, Omary MB. Type II keratins are phosphorylated on a unique motif during stress and mitosis in tissues and cultured cells. Molecular Biology Cell. 2002;**13**(6):1857-1880

[4] Busch T, Armacki M, Eiseler T, Joodi G, Temme C, Jansen J, et al. Keratin 8 phosphorylation regulates keratin reorganization and migration of epithelial tumor cells. Journal Cell Science. 2012;**125**(Pt 9): 2148-2159

[5] Lane EB, Goodman SL, Trejdosiewicz LK. Disruption of the keratin filament network during epithelial cell division. EMBO Journal. 1982;**1**(11):1365-1372

[6] Seltmann K, Fritsch AW, Käsb JA, Magin TM. Keratins significantly contribute to cell stiffness and impact invasive behavior. PNAS USA. 2013; **110**(46):18507-18512

[7] Paladini RD, Takahashi K, Bravo NS, Coulombe PA. Onset of re-epithelialization after skin injury correlates with a reorganization of keratin filaments in wound edge keratinocytes: Defining a potential role for keratin 16. Journal Cell Biology. 1996;**132**(3):381-397

[8] Mazzalupo S, Wong P, Martin P, Coulombe PA. Role for keratins 6 and 17 during wound closure in embryonic mouse skin. Developmental Dynamics. 2003;**226**(2):356-365

[9] Dmello C, Srivastava SS, Tiwari R, Chaudhari PR, Sawant S, Vaidya MM. Multifaceted role of keratins in epithelial cell differentiation and transformation. Journal Biosciences. 2019;**44**(2):33

[10] Homberg M, Magin TM. Beyond expectations: Novel insights into epidermal keratin function and regulation. International Review Cell Molecular Biology. 2014;**311**(Chapter 6): 265-306

[11] Yang L, Zhang S, Wang G. Keratin 17 in disease pathogenesis: From cancer to dermatoses. Journal Pathology. 2018; **247**(2):158-165. DOI: 10.1002/path.5178

[12] Liu J, Liu L, Cao L, Wen Q. Keratin 17 promotes lung adenocarcinoma progression by enhancing cell proliferation and invasion. Medical Science Monitor. 2018;**24**:CLR4782-4790. DOI: 10.12659/MSM.909350

[13] Bernerd F, Magnaldo T, Freedberg IM, Blumenberg M. Expression of the carcinoma-associated keratin K6 and the role of AP-1 proto-oncoproteins. Gene Expression. 1993; **3**(2):187-199

[14] Jacob JT, Nair RR, Poll BG, Pineda CM, Hobbs RP, Matunis MJ, et al. Keratin 17 regulates nuclear morphology and chromatin organization. Journal Cell Science. 2020; **133**(20):jcs254094

[15] Song D-G, Kim YS, Jung BC, Rhee K-J, Pan C-H. Parkin induces upregulation of 40S ribosomal protein SA and posttranslational modification of cytokeratins 8 and 18 in human cervical cancer cells. Applied Biochemistry Biotechnology. 2013;**171**: 1630-1638

[16] Liao J, Omary MB. 14-3-3 proteins associate with phosphorylated simple

epithelial keratins during cell cycle progression and act as a solubility cofactor. Journal Cell Biology. 1996; **133**(2):345-357

[17] Mariani RA, Paranjpe S, Dobrowolski R, Weber GF. 14-3-3 targets keratin intermediate filaments to mechanically sensitive cell–cell contacts. Molecular Biology Cell. 2020;**31**(9): 930-943

[18] Fortier A-M, Asselin E, Cadrin M. Keratin 8 and 18 loss in epithelial cancer cells increases collective cell migration and cisplatin sensitivity through claudin1 up-regulation. Journal Biological Chemistry. 2013;**288**(16): 11555-11571

[19] Chu PG, Weiss LM. Keratin expression in human tissues and neoplasms. Histopathology. 2002;**40**(5): 403-439

[20] McDowell EM, Keenan KP, Huang M. Effects of vitamin A-deprivation on hamster tracheal epithelium: A quantitative morphologic study. Virchows Archiv B Cell Pathologie. 1984;**45**(2):197-219

[21] Huang FL, Roop DR, De Luca LM. Vitamin A deficiency and keratin biosynthesis in cultured hamster trachea. In Vitro Cellular Developmental Biology. 1986;**22**(4): 223-230

[22] Leube RE, Rustad TJ. Squamous cell metaplasia in the human lung: Molecular characteristics of epithelial stratification. Virchows Archiv B Cell pathology including molecular mathology. 1991;**61**(4):227-253

[23] Jetten AM, George MA, Smits HL, Vollberg TM. Keratin 13 expression is linked to squamous differentiation in rabbit tracheal epithelial cells and down-regulated by retinoic acid. Experimental Cell Research. 1989; **182**(2):622-634

[24] Kaartinen L, Nettesheim P, Adler KB, Randell SH. Rat tracheal epithelial cell differentiation in vitro. In Vitro Cellular Developmental Biology. 1993;**29A**(6):481-492

[25] Schlage WK, Bülles H, Friedrichs D, Kuhn M, Teredesai A, Terpstra PM. Cytokeratin expression patterns in the rat respiratory tract as markers of epithelial differentiation in inhalation toxicology. II. Changes in cytokeratin expression patterns following 8-day exposure to room-aged cigarette sidestream smoke. Toxicologic Pathology. 1998;**26**(3):344-360

[26] Jetten AM. Multistep process of squamous differentiation in tracheobronchial epithelial cells in vitro: Analogy with epidermal differentiation. Environmental Health Perspectives. 1989;**80**:149-160

[27] Wang GF, Lai MD, Yang RR, Chen PH, Su YY, Lv BJ, et al. Histological types and significance of bronchial epithelial dysplasia. Modern Pathology. 2006;**19**(3):429-437

[28] Broers JL, Ramaekers FC, Rot MK, Oostendorp T, Huysmans A, van Muijen GN, et al. Cytokeratins in different types of human lung cancer as monitored by chain-specific monoclonal antibodies. Cancer Research. 1988; **48**(11):3221-3229

[29] Chen Y, Cui T, Yang L, Mireskandari M, Knoesel T, Zhang Q, et al. The diagnostic value of cytokeratin 5/6, 14, 17, and 18 expression in human non-small cell lung cancer. Oncology. 2011;**80**(5–6):333-340

[30] Blobel GA, Moll R, Franke WW, Vogt-Moykopf I. Cytokeratins in normal lung and lung carcinomas I. adenocarcinomas, squamous cell carcinomas and cultured cell lines. Virchows Archiv B Cell pathology including molecular pathology. 1984; **45**(4):407-429

[31] Manger RL, Heckman CA. Structural anomalies of highly malignant respiratory tract epithelial cells. Cancer Research. 1982;**42**(11):4591-4599

[32] Heckman CA. Organ- and species-specificity of epithelial growth. In Vitro. 1983;**19**(1):31-40

[33] Marchok AC, Rhoton JC, Griesemer RA, Nettesheim P. Increased *in vitro* growth capacity of tracheal epithelium exposed *in vivo* to 7,12-dimethylbenz(a)anthracene. Cancer Research. 1977;**37**(6):1811-1821

[34] Marchok AC, Rhoton JC, Nettesheim P. *In vitro* development of oncogenicity in cell lines established from tracheal epithelium preexposed *in vivo* to 7,12-dimethylbenz(a) anthracene. Cancer Research. 1978; **38**(7):2030-2037

[35] Heckman CA, Olson AC. Morphological markers of oncogenic transformation in respiratory tract epithelial cells. Cancer Research. 1979; **39**(7):2390-2399

[36] Steele VE, Marchok AC, Nettesheim P. Establishment of epithelial cell lines following exposure of cultured tracheal epithelium to 12-O-tetradecanoyl-phorbol-13-acetate. Cancer Research. 1978;**38**(10):3563-3565

[37] Jamasbi RJ, Nettesheim P, Kennel SJ. Demonstration of cellular and humoral immunity to transplantable carcinomas derived from respiratory tracts of rats. Cancer Research. 1978;**38**(2):261-267

[38] Papanicolaou ON. Atlas of Exfoliative Cytology. Cambridge, MA: Harvard University Press; 1954. p. 202

[39] Franke WW, Mayer D, Schmid E, Denk H, Borenfreund E. Differences of expression of cytoskeletal proteins in cultured rat hepatocytes and hepatoma cells. Experimental Cell Research. 1981; **134**(2):345-365

[40] O'Farrell PH. High resolution two dimensional electrophoresis of proteins. Journal Biological Chemistry. 1975; **250**(10):4007-4021

[41] Ridder G, VonBargen E, Burgard D, Pickrum H, Williams E. Quantitative analysis and pattern recognition of two-dimensional electrophoresis gels. Clinical Chemistry. 1984;**30**(12 Pt 1):1919-1924

[42] Li R, Hao J, Fujiwara H, Xu M, Yang S, Dai S, et al. Analytical characterization of methyl-β-cyclodextrin for pharmacological activity to reduce lysosomal cholesterol accumulation in Niemann-Pick disease type C1 cells. ASSAY Drug Development Technologies. 2017;**15**(4):154-166

[43] Prot pi Protein Tool. Available from: https://www.protpi.ch/Calculator [Accessed: December 20, 2020]

[44] ImageJ. Available from: https://imagej.net/ImageJ1987 [Accessed: December 20, 2020]

[45] Schlage WK, Bülles H, Friedrichs D, Kuhn M, Teredesai A. Cytokeratin expression patterns in the rat respiratory tract as markers of epithelial differentiation in inhalation toxicology. I. Determination of normal cytokeratin expression patterns in nose, larynx, trachea, and lung. Toxicologic Pathology. 1998;**26**(3):324-343

[46] Bragulla HH, Homberger DG. Structure and functions of keratin proteins in simple, stratified, keratinized and cornified epithelia. Journal Anatomy. 2009;**214**(4):516-559

[47] Moll R, Franke WW, Schiller DL, Geiger B, Krepler R. The catalog of human cytokeratins: Patterns of expression in normal epithelia, tumors and cultured cells. Cell. 1982;**31**(1):1-24

[48] Moll R. Cytokeratine Als Differenzierungsmarker. Expressionsprofile von Epithelien Und

Epithelialen Tumoren [Cytokeratins as Markers of Differentiation. Expression Profiles in Epithelia and Epithelial Tumors]. New York: Gustav Fisher Verlag; 1993. pp. 1-197

[49] Foster MW, Gwinn WM, Kelly FL, Brass DM, Valente AM, Moseley MA, et al. Proteomic analysis of primary human airway epithelial cells exposed to the respiratory toxicant diacetyl. Journal Proteome Research. 2017;**16**(2):538-549

[50] Hackett NR, Shaykhiev R, Walters MS, Wang R, Zwick RK, Ferris B, et al. The human airway epithelial basal cell transcriptome. PLoS One. 2011;**6**(5):e18378

[51] Gharib TG, Chen G, Wang H, Huang C-C, Prescott MS, Shedden K, et al. Proteomic analysis of cytokeratin isoforms uncovers association with survival in lung adenocarcinoma. Neoplasia. 2002;**4**(5):440-448

[52] Cole BB, Smith RW, Jenkins KM, Graham BB, Reynolds PR, Reynolds SD. Tracheal basal cells: A facultative progenitor cell pool. American Journal Pathology. 2010;**177**(1):362-376

[53] Nakajima M, Kawanami O, Jin E, Ghazizadeh M, Honda M, Asano G, et al. Immunohistochemical and ultrastructural studies of basal cells, Clara cells and bronchiolar cuboidal cells in normal human airways. Pathology International. 1998;**48**(12):944-953

[54] Terzaghi M, Nettesheim P, Williams ML. Repopulation of denuded tracheal grafts with normal, preneoplastic, and neoplastic epithelial cell populations. Cancer Research. 1978;**38**:4546-4553

[55] Wang F, Zieman A, Coulombe PA. Skin keratins. Methods Enzymology, "Intermediate Filament Proteins". 2016; **568**:303-350

[56] Tong X, Coulombe PA. A novel mouse type I intermediate filament

gene, keratin 17n (K17n), exhibits preferred expression in nail tissue. Journal Investigative Dermatology. 2004;**122**(4):965-970

[57] Ooi AT, Mah V, Nickerson DW, Gilbert JL, Ha VL, E. A, et al. Presence of a putative tumor-initiating progenitor cell population predicts poor prognosis in smokers with non-small cell lung cancer. Cancer Research. 2010;**70**(16): 6639-6648

[58] Smirnova NF, Schamberger AC, Nayakanti S, Hatz R, Behr J, Eickelberg O. Detection and quantification of epithelial progenitor cell populations in human healthy and IPF lungs. Respiratory Research. 2016;**17**:83

[59] Heckman CA, Marchok AC, Nettesheim P. Respiratory tract epithelium in primary culture: Concurrent growth and differentiation during establishment. Journal Cell Science. 1978;**32**:269-291

[60] Nagashio R, Sato Y, Matsumoto T, Kageyama T, Satoh Y, Ryuge S, et al. Significant high expression of cytokeratins 7, 8, 18, 19 in pulmonary large cell neuroendocrine carcinomas, compared to small cell lung carcinomas. Pathology International. 2010;**60**(2): 71-77

[61] Rogel MR, Jaitovich A, Ridge KM. The role of the ubiquitin proteasome pathway in keratin intermediate filament protein degradation. Proceedings American Thoracic Society. 2010;**7**(1):71-76

[62] Bär J, Kumar V, Roth W, Richter M, Leube RE, Magin TM. Skin fragility and impaired desmosomal adhesion in mice lacking all keratins. Journal Investigative Dermatology. 2014;**134**(4):1012-1022

[63] Lin A, Li S, Feng C, Yang S, Wang H, Ma D, et al. Stabilizing mutations of KLHL24 ubiquitin ligase cause loss of keratin 14 and human skin

fragility. Nature Genetics. 2016;**48**(12): 1508-1516

[64] Jaitovich A, Mehta S, Ciechanover A, Goldman RD, Ridge KM. Ubiquitin-proteasome-mediated degradation of keratin intermediate filaments in mechanically stimulated A549 cells. Journal Biological Chemistry. 2008;**283**(37):25348-25355

[65] Kanaji N, Bandoh S, Ishii T, Fujita J, Ishida T, Matsunaga T, et al. Cytokeratins negatively regulate the invasive potential of lung cancer cell lines. Oncology Reports. 2011;**26**(4): 763-768

[66] Stosiek P, Kasper M, Moll R. Changes in cytokeratin expression accompany squamous metaplasia of the human respiratory epithelium. Virchows Archiv A Pathological Anatomy. 1992;**421**(2):133-141

[67] Vijayaraj P, Kröger C, Reuter U, Windoffer R, Leube RE. Magin TM keratins regulate protein biosynthesis through localization of GLUT1 and −3 upstream of AMP kinase and raptor. Journal Cell Biology. 2009;**187**(2): 175-184

[68] Rotty JD, Coulombe PA. A wound-induced keratin inhibits Src activity during keratinocyte migration and tissue repair. Journal Cell Biology. 2012; **197**(3):381-389

Chapter 2

Genetic Abnormalities, Melanosomal Transfer, and Degradation inside Keratinocytes Affect Skin Pigmentation

Md. Razib Hossain, Miho Kimura-Sashikawa and Mayumi Komine

Abstract

Skin pigmentation is a specific and complex mechanism that occurs as a result of the quantity and quality of melanin produced, as well as the size, number, composition, mode of transfer, distribution, and degradation of the melanosomes inside keratinocytes and the handling of the melanin product by the keratinocyte consumer. Melanocyte numbers typically remain relatively constant. Melanin synthesis, melanosome maturation, and melanoblast translocation are considered to be responsible for hereditary pigmentary disorders. Keratinocytes play a significant role in regulating the adhesion, proliferation, survival, and morphology of melanocytes. In the epidermis, each melanocyte is surrounded by 30–40 keratinocytes through dendrites and transfers mature melanosomes into the cytoplasm of keratinocytes, which are then digested. Melanocytes are believed to transfer melanosomes to neighboring keratinocytes via exocytosis-endocytosis, microvesicle shedding, phagocytosis, or the fusion of the plasma membrane, protecting skin cells against ultraviolet (UV) damage by creating a physical barrier (cap structure) over the nucleus. An understanding of the factors of melanocytes and keratinocytes that induce pigmentation and the transfer mechanism of melanosomes to keratinocytes and how genetic abnormalities in keratinocytes affect pigmentary skin disorders will help us to elucidate hereditary pigmentary disorders more transparently and provide a conceptual framework for the importance of keratinocytes in the case of pigmentary disorders.

Keywords: melanin transfer, melanosome, melanocytes, keratin, keratinocytes, skin pigmentation

1. Introduction

The skin is the outermost organ, covering the whole body. It helps with temperature regulation, immune defense, vitamin production, and sensation. However, skin is also associated with many potential problems, with over 3,000 possible disorders [1].

Furthermore, its color has social, spiritual, cosmetic, and medical issues associated with it. The color of human skin depends on the distribution of melanin, a pigment produced in melanosomes in the melanocyte cytoplasm via the tyrosinase reaction. Skin produces melanin within the melanocytes of the interfollicular epidermis through a multi-stage process called melanogenesis. The subsequent transfer, translocation, and degradation of melanin to, in, and by the recipient keratinocytes, respectively, causes pigmentation. The skin can protect itself from solar irradiation, thanks to this specific and complex pigmentation mechanism. The nature of pigmentation is determined by the quantity and quality (pheo/eumelanin ratio) of melanin and the size, number, composition, mode of transfer, distribution, and degradation of the melanosomes inside keratinocytes, as well as the handling of the melanin product by macrophages. On the other hand, melanocyte numbers tend to remain relatively constant. The dysregulation of the process of melanogenesis can cause several types of pigmentary defects, classified as either hypopigmentation, hyperpigmentation, or mixed hyper/hypopigmentation [2–4]. Hereditary pigmentary disorders occur mostly due to genetic deficiency in melanin, irregular melanin synthesis in melanocytes, abnormal melanosome maturation, and melanoblast translocation. Various physiological factors, including autocrine and paracrine hormones/cytokines, also modulate skin pigmentation.

Keratins are specific to epithelial cells and one of the cytoskeletal proteins that provide structural support to keratinocytes through intermediate filament networks. Approximately 21 different keratins have been reported to be associated with different hereditary disorders [5]. Keratinocytes are crucial in the organization of cell adhesion, as well as the proliferation, survival, and morphology of melanocytes. Keratins also play a pivotal role in the uptake of melanosomes into keratinocytes, organelle transport, and nuclear anchorage, indicative of their involvement in intracellular transportation [6]. Several studies have claimed that the keratin 5 head domain interacts with heat shock cognate 70 (Hsc70) and is involved in organelle transport [7, 8] and chaperone-mediated autophagy.

Many studies have been performed to elucidate the role of keratinocytes in pigmentation; however, the transfer mechanism of melanin to keratinocytes remains ambiguous. It has been postulated that melanocytes transfer melanosomes to neighboring keratinocytes via exocytosis-endocytosis, microvesicle shedding, or phagocytosis. In this chapter, the factors of melanocytes and keratinocytes that induce pigmentation and the potential mechanism of melanosomal transportation to the surrounding keratinocytes, and how genetic abnormalities in keratinocytes affect pigmentary skin disorders are reviewed to develop a basis for the role of keratinocytes in pigmentation.

2. Skin pigmentation is regulated by several factors of melanocytes

2.1 Melanosome biogenesis and melanogenesis process

Melanocytes are unique neural crest-derived cells that synthesize and store melanin pigments in melanosomes, which are specific membrane-bound organelles that share several features with lysosomes. In particular, they contain acid-dependent hydrolases and lysosomal-associated membrane proteins [9]. Melanosomes are members of the cell-specific organelle family, termed lysosome-related organelles (LROs), which also comprise lytic granules observed in cytotoxic T lymphocytes and natural killer cells, MHC class II compartments (MIICs) observed in antigen-presenting cells, basophil granules, platelet-dense granules, azurophil granules observed in neutrophils, and Weibel-Palade bodies observed in endothelial cells [10].

Most pigment-specific proteins are localized in melanosomes [11] and are divided into three distinct groups—structural fibrillar proteins required for melanosome structure and binding of melanin, enzymatic components required for melanin synthesis, and proteins required for melanosome transport and distribution [12, 13].

Melanosomes can be morphologically classified into four distinct stages (I–IV) based on melanosomal maturation and three important structural proteins that form melanosomes—melanosomal matrix protein (PMEL17/Silv/GP100), melanoma antigen recognized by T cell-1 (MART-1), and glycoprotein nonmetastatic melanoma protein b (GPNMB/DC-HIL/osteoactivin) (**Figure 1**). Intraluminal fibrils begin to form in amorphous spherical stage I melanosomes and develop a meshwork characteristic of stage II melanosomes, both of which are considered to be pre-melanosomes and do not contain melanin. MART-1 has been previously observed in earlier melanosomal stages [14]. In stage II, melanin synthesis begins within the fibrillar and is deposited uniformly on the internal fibrils that evolve into stage III. Melanin is deposited on the PMEL17 fibrils in stage III. MART-1 plays a significant role in the maturation of PMEL17 [14]. In the last stage (IV) of maturation, copious amounts of melanin fill the melanosomes and form a masked internal structure and dark color. GPNMB, a melanosome-specific and proteolytically released protein, is superabundant in late melanosomes [15] and is crucial for the formation of melanosomes in a microphthalmia transcription factor (MITF)-independent fashion [16]. Recent studies have shown that early melanosomes are derived from the endoplasmic reticulum (ER), coated vesicles, lysosomes, and endosomes [17–19].

The enzymatic components of melanosomes help melanosomes reach their ultimate stage of melanosomal maturation. Melanosomal enzymatic components, including TYR, tyrosinase-related protein-1 (TYRP1), and dopachrome tautomerase/tyrosinase-related protein-2 (DCT/TYRP2), play major roles in melanin synthesis (**Figure 1**). TYRP, a critical copper-dependent enzyme, catalyzes the conversion of L-tyrosine to L-3,4-dihydroxyphenylalanine (L-DOPA), the rate-limiting step in melanin synthesis. Copper oxidization deactivates this enzyme but can be activated by electron donors, such as L-DOPA, ascorbic acid, superoxide anion, and nitric

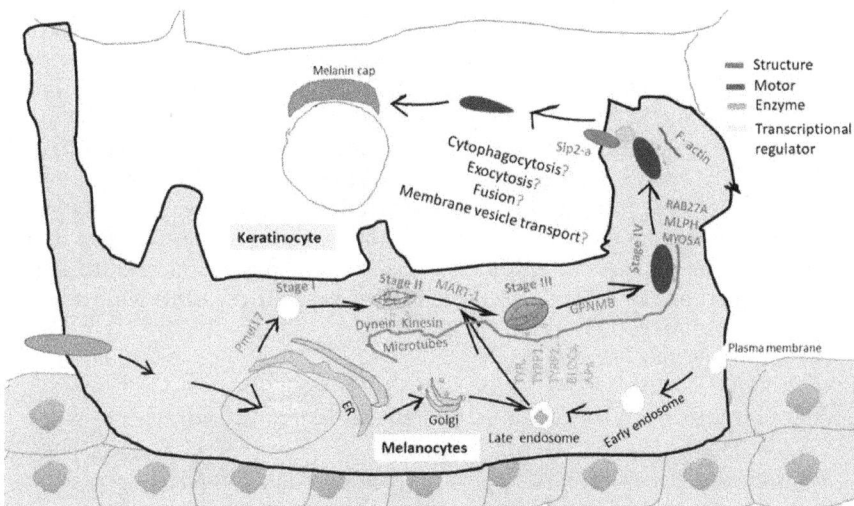

Figure 1.
Factors of melanocyte that regulate skin pigmentation. The figure illustrates components that are involved in melanosomal transportation, enzymatic activates, motor activates, and transcriptional regulation.

oxide (NO) [3, 20, 21]. Protein kinase C-β (PKC-β) phosphorylates two serine residues of the cytoplasmic domain and activates tyrosinase [22]. Mutations that inactivate this enzyme result in the most severe form of oculocutaneous albinism (OCAIa). TYRP1 and TYRP2 are also found in the membrane of melanosomes, and it is assumed that TYRP1 is involved in the activation and stabilization of tyrosinase, melanosome synthesis, increasing the eumelanin/pheomelanin ratio, and working against oxidative stress due to its peroxidase effect [3, 20].

2.2 Melanosome trafficking

Melanosomes move from the perinuclear area to the periphery of melanocytes, and melanocytes transfer packaged melanin into adjacent keratinocytes because of the function of microtubules, actin filaments, and myosin, resulting in skin pigmentation [13]. Early melanosomes originate in the perinuclear area and move toward the periphery of melanocytes (i.e., dendrites) by kinesin and dynein mediate microtubule-dependent intracellular transportation systems (**Figure 1**). During this period, they eventually mature and turn into late (pigmented) melanosomes [23, 24]. Kinesins ensure the centrifugal movement of melanosomes, and melanosomal cargo is transferred from microtubules to F-actin in dendrites. RAB27A, melanophilin (MLPH), and myosin-Va (MYO5A) induce complexes to connect melanosomes to F-actin-based motors. Mutations in any of these genes induce a noticeable accumulation of pigments in the perinuclear region of mutant melanocytes due to the disruption of their transport to the dendrites [25], resulting in various forms of Griscelli syndrome (types II, III, and I, respectively) in humans [26]. This is manifested by mouse Melan-a cells by RAB27A linking to synaptotagmin-like 2 (SYTL2), prompting SYTL2 to dock melanosomes at the plasma membrane, suggesting that SYTL2 plays a role as a regulator of melanosome exocytosis [27–29].

2.3 Melanogenic regulation in melanocytes

The most vital transcription factor that regulates melanocyte function is MITF, which controls the expression of the melanogenesis enzymes TYR, TYRP1, and TYRP2 (**Figure 1**) [30]. Mutations in MITF result in Waardenburg syndrome type 2 (WS2) [31]. The MITF promoter is regulated by various other transcription factors, including a paired box protein 3 (PAX3), sex-determining region Y-box 9 and 10 (SOX9 and SOX10), lymphoid enhancer-binding factor 1 (LEF-1), and cyclic adenosine monophosphate (cAMP) responsive element-binding protein (CREB), which is phosphorylated by signals via the melanocortin-1 receptor (MC1R) [13]. The roles of polymorphisms in MC1R have been thoroughly investigated in response to UV radiation and/or in controlling constitutive skin pigmentation among racial/ethnic groups [32]. Several physiological factors from fibroblasts, keratinocytes, and other sources also regulate the expression levels and functions of MITF [33].

3. Skin pigmentation is regulated by several factors of keratinocytes

Keratinocytes derived from dark skin, such as microphthalmia-associated transcription factors and tyrosinase, significantly stimulate the expression of melanocyte-specific proteins. It has been suggested that keratinocytes regulate skin pigmentation, at least in part, regardless of whether the melanocytes are derived from light or dark skin [34, 35].

3.1 Keratinocyte-derived factors regulate melanocytes

Currently, it has been postulated that Foxn1/Whn/Hfh11, a transcription factor expressed by keratinocytes, is a regulator of keratinocyte growth and differentiation, and is also involved in melanocyte recruitment and induction of pigmentation in the skin through basic fibroblast growth factor (bFGF) production [36]. In summary, Foxn1 recruits melanocytes to their desired position and induces melanosome transfer, acting as an activator of the pigment-recipient phenotype. Keratinocyte-derived factors that act as activators of melanocytes also include stem cell factor (SCF), hepatocyte growth factor (HGF), granulocyte-macrophage colony-stimulating factor (GM-CSF), nerve growth factor (NGF), α-melanocyte-stimulating hormone (α-MSH), adrenocorticotropic hormone (ACTH), endorphin, endothelin-1 (ET-1), prostaglandin (PG)E2/PGF2a, and leukemia inhibitory factor (LIF) (**Figure 2**) [37]. However, whether any of these factors are regulated by Foxn1 is unclear.

3.2 Melanosome transfer

Synapses between melanocytes and keratinocytes are believed to exist, such as in the neural system, wherein melanosome transfer occurs through these synapses via some unknown mechanisms. Protease (proteinase)-activated receptor-2 (PAR-2), a G-protein coupled receptor, which is expressed on keratinocytes, seems to be closely involved in melanosome transfer. PAR-2 has a crucial role in mediating the phagocytosis of melanosomes in a Rho-dependent manner and in determining skin color phenotype [38]. Recently, it has been shown that the keratinocyte growth factor receptor (KGFR) plays a role similar to PAR-2 [39]. Keratinocyte phagocytosis of latex beads is enhanced by KGFR activation, and the addition of KGF to co-cultures induces the transfer of tyrosinase-positive granules. Phagocytosis via KGFR is dependent on the PAR-2-Rho pathway, as well as on Rac and Cdc42 activation [39]. Apart from phagocytosis, PAR-2 stimulates melanocyte dendricity and contributes to skin pigmentation. Stimulated keratinocytes release prostaglandins, PGE2, and PGF2a, which bind to the surface of melanocytes, thereby inducing dendrite

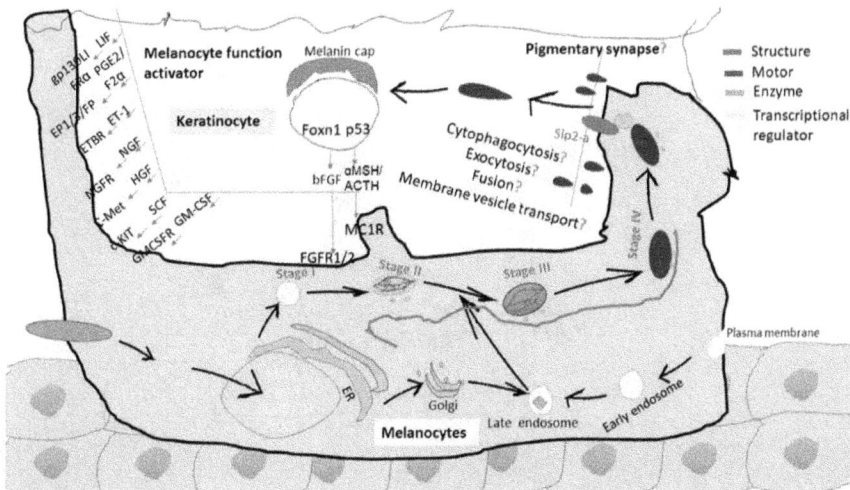

Figure 2.
Factors of Keratinocyte that regulate skin pigmentation. Foxn1 and p53 are directly involved in the up-regulation of pigmentation through bFGF and POMC derivatives, receptively. The figure shows all the relevant factors that enhance melanocyte growth and function.

formation [40]. Several physiological factors regulate the expression of PAR-2 (38), such as dickkopf 1 (DKK1), an inhibitor of the Wnt/b-catenin pathway [41].

The process of melanosome transfer to keratinocytes is not completely understood at this time, although various assumptions have been proposed, including exocytosis, cytophagocytosis, fusion, and membrane vesicle transport [26].

4. Proposed mechanism of melanosomal transfer to keratinocytes

4.1 Exo/endocytosis

Regulated exocytosis is a multistage process in which the membranes of cytoplasmic organelles fuse with the plasma membrane in response to stimulation through which secretions are released from the vesicle to the cell exterior. Several types of regulated secretory exocytosis exist, including the exocytosis of synaptic vesicles and dense-core vesicles in the presynaptic compartment of neuronal cells, exocytosis of these vesicles (all over the plasma membrane) by neuroendocrine and endocrine cells, exocytosis of secretory granules at the apical plasma membrane of exocrine cells, and exocytosis by hematopoietic cells of various types of secretory organelles (**Figure 3**) [42].

According to the exocytosis theory, the melanosomal membrane fuses with the melanocyte plasma membrane, resulting in extracellular melanin, and the melanin

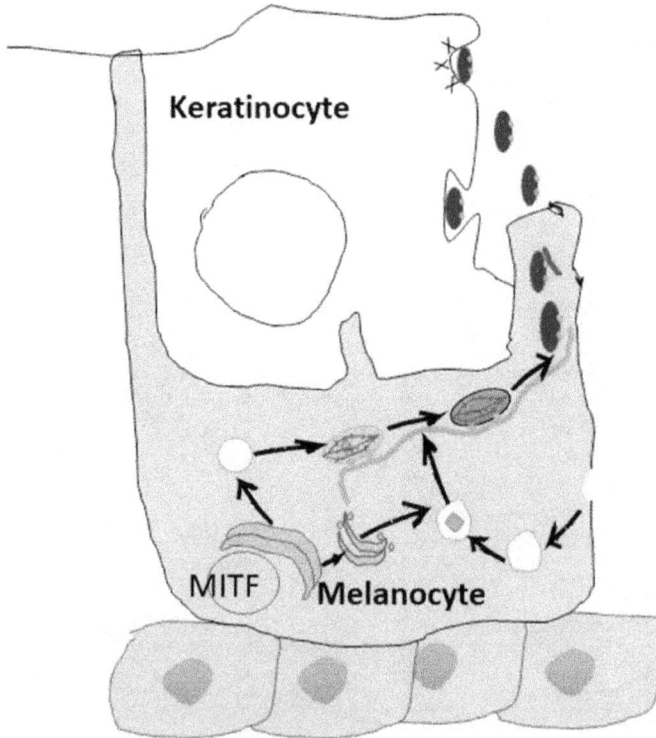

Figure 3.
Exocytosis as a mode of melanosomal transfer to the keratinocyte. Melanosomal membrane fuses with the plasma membrane, thus release melanin. This released melanin is taken up by keratinocytes either endocytosis or phagocytosis.

that is released is phagocytosed by a neighboring keratinocyte. As a result, melanin is transferred from melanocytes to keratinocytes (**Figure 3**).

This hypothesis is based on observations of human skin and hair follicles using an electron microscope, depicting ranased is phagocytosed by a neighboring keratinocyte. As a result, melanin is trannin is tranratinocyte. As a resultclathrin-coated pits [43, 44].

Melanocytes release melanin in the extracellular space *in vitro* after being stimulated either with α-melanocyte-stimulating hormone or with the soluble domain of the β-amyloid precursor protein [45, 46]. The process is referred to as exocytosis because the released melanin is not surrounded by a membrane, as observed by electron microscopy, and melanocytes are known to express SNAREs and Rab GTPases molecules, which are involved in the regulation of exocytosis.

SNAREs contain three conserved families of membrane-associated proteins (synaptobrevin/VAMP, syntaxin, and SNAP25 families) that play a significant role in the later stages of membrane fusion. Several SNAREs have been identified in melanosome-enriched fractions, including SNAP23, SNAP25, VAMP2, syntaxin 4, and syntaxin 6 [47, 48]. SNAP25 and syntaxin on the plasma membrane bind VAMP to the vesicle membrane. Some immunoprecipitation experiments have reported a degree of interaction between VAMP2 and SNAP23, but not with syntaxin 4, to achieve fusion.

Rab GTPases are another family of proteins that act in the tethering and docking of membranes before fusion. RAB27A contributes to the regulated exocytosis of various types of organelles [27]. Synaptotagmin-like protein2-a (Slp2-a) is a concurrent RAB27A effector that has been found in melanocytes [27]. It links RAB27A with phosphatidylserine and facilitates the attachment of melanosomes to the plasma membrane, which is an essential step in exocytosis. Interestingly, Slp2-a is a putative regulator of exocytosis at the neuronal synapse and is structurally homologous to synaptotagmin. Coupled exo/endocytosis of melanin transfer from melanocytes to keratinocytes in the core of the melanosome, termed melanocore [49] and the small GTPase Rab11b mediates the final steps of melanocore exocytosis prior to the transfer to keratinocytes [49]. One study found that the depletion of Rab11b, but not RAB27A, causes a decrease in keratinocyte-induced melanin exocytosis [49]. The exocyst is an evolutionarily conserved protein complex comprising eight subunits, including Sec8, Sec15, and Exo70 [50]. This complex plays an essential role in various processes, including cell migration, vesicle tethering, membrane trafficking, ciliogenesis, autophagy, and cytokinesis. It is postulated that the exocyst is essential for melanocyte exocytosis and keratinocyte transfer [51]. Rab3, consisting of Rab3a-d, is also involved in exocytosis in several cell types [52, 53]. Rab3a is expressed in melanocytes, and its expression is downregulated by UV irradiation [47, 54].

Further support for the exocytosis hypothesis is based on findings regarding melanin in the keratinocyte cytoplasm, which are present in the same way it exogenously administered melanin by melanocytes [45, 55–58]. The distribution pattern of melanin granules seems to be dependent on size. The phagocytosis of small or large latex beads also results in aggregates and singly dispersed beads, respectively. Melanin granules aggregate after being ingested as single granules, indicative of the final stage in the lifecycle of melanin in keratinocytes.

Finally, melanocytes are closely related to both neuronal and hematopoietic cells because of their neural crest origin and secretory lysosome family belonging, respectively [59]. Both synaptic vesicles and secretory lysosomes help regulate exocytosis upon stimulation, which suggests that melanin transfer occurs via similar mechanisms.

4.2 Cytophagocytosis

Phagocytosis is an essential process that maintains cellular homeostasis and is defined as the cellular engulfment of particles with a diameter of more than 0.5

mm. In the epidermis, the phagocytosis of melanosomes into keratinocytes is vital to protect their DNA against damage from ultraviolet B (UVB) radiation and is essential for triggering host defenses against invading pathogens, as well as in the elimination of damaged, senescent, and apoptotic cells in mammals. Phagocytes can be classified into two types—professional, such as macrophages, dendritic cells, and granulocytes and non-professional (or amateur), such as keratinocytes. Amateur phagocytes are slower, less mobile, and have a limited range of particles that they can take up [60]. Their phagocytic nature has been shown both in vitro [55] and in vivo [56].

The phagocytosis of a viable cell or an intact part of a viable cell is known as cytophagocytosis. The cytophagocytosis hypothesis of melanin transfer denotes the phagocytosis of intact melanocytic dendrite cells, known as the "dendrite tip." First, the melanocytes extend their dendrites towards the surrounding keratinocytes to make contact. The keratinocytes respond with extensive membrane ruffling and engulf the dendrite tip using villus-like cytoplasmic projections. Next, the dendrite tip is squeezed and pinched off, thus forming a cytoplasmic poach filled with melanosomes. Then, a phagolysosome is formed by the fusion of lysosomes, and the degradation of the melanocyte membranes and cytoplasmic constituents occurs. Meanwhile, phagolysosomes are transported to the supranuclear region. Finally, the phagolysosome disintegrates into smaller vesicles containing a single melanin granule or aggregates of melanin granules, and are then dispersed over the cytoplasm (**Figure 4**) [61].

The hypothesis of the cytophagocytosis of melanosomes by keratinocytes is supported by evidence obtained using electron microscopy [43, 62–64]. Additionally, measuring the internalization of latex microsphere beads by keratinocyte phagocytosis has been shown to be activated by keratinocyte growth factor, which acts only on the recipient keratinocytes [39, 45]. The activation or inhibition of PAR-2 expressed by keratinocytes, but not by melanocytes [65], regulates melanosome transfer via keratinocyte phagocytosis [39, 66]. Light and electron microscopy showed that exogenously added melanosomes are taken up by normal human keratinocytes in a time-dependent manner, reflecting a possible melanosome transfer process in which melanosomes released into the extracellular space are phagocytosed by keratinocytes [67].

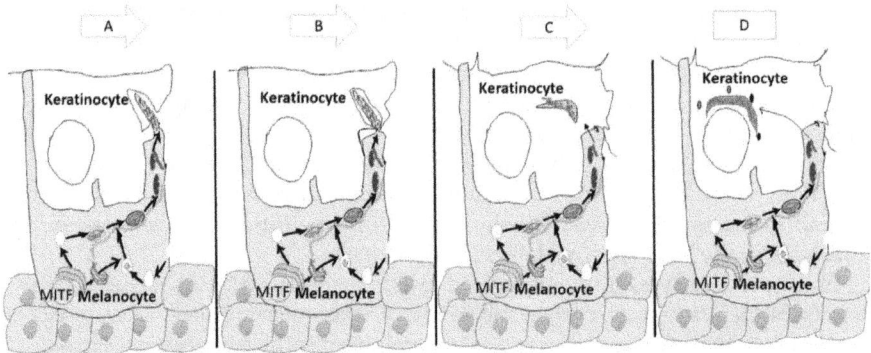

Figure 4.
Cytophagocytosis as a mode of melanosomal transfer to the keratinocyte. In this process, melanocytic dendrite is pinched off and phagocytosed, and resulting in the phagolysosome. Melanin granules are dispersed throughout the cytoplasm from this phagolysosome.

4.3 Fusion of plasma membranes as a mode of transfer

The melanocyte plasma membrane fuses with the keratinocyte plasma membrane, thus creating a pore or channel that connects the cytoplasm of both cells and through which melanosomes are transported (**Figure 5**).

The fusion assumption of pigment transfer has been suggested in pigmented basal cell carcinoma and the skin of black guinea pig ears [68]. Recently, it has been suggested that filopodia extend from the dendrite tips and cell body of melanocytes and fuse with the neighboring keratinocyte membrane, allowing for the passage of melanosomes [69] towards the keratinocyte membrane. Although the melanosomal transfer was observed via these protrusions, proof of membrane fusion was not unambiguously obtained. The similar optical properties of melanocyte and keratinocyte membranes make it difficult to distinguish fusion via light microscopy. In contrast, thin projections were found directly connecting the melanocyte and keratinocyte cytoplasm in the electron micrographs of their co-cultures.

This mode of cell-cell communication transport could be considered as tunneling nanotubes [69], providing a network of various cultured cells and functioning as channels for organelle transport [70].

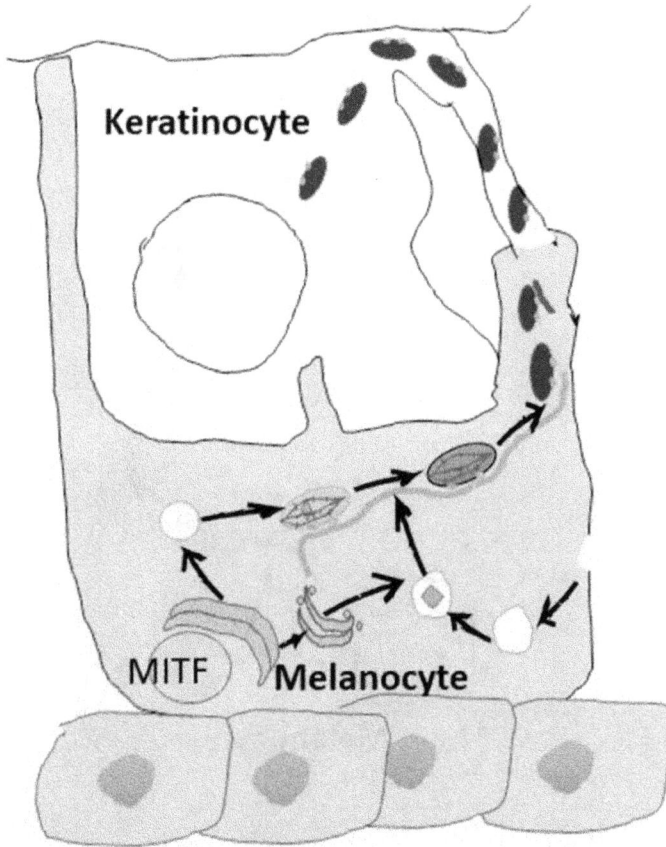

Figure 5.
Fusion as a mode of melanosomal transfer to the keratinocyte. Melanosomes are passage through a channel. The plasma membrane of the melanocyte fuses with the plasma membrane of Keratinocyte to form this channel.

Filopodial fusion with neighboring cells forms a tubular structure composed of actin filaments, with a diameter of 50–200 nm, directly connecting the cytoplasm of cells. The tube allows for the unidirectional transport of organelles and plasma membrane molecules, as opposed to soluble cytoplasmic molecules.

Similarly, interconnecting channels have been observed between cytotoxic T lymphocytes or natural killer cells, antigen-presenting cells to B cells and their respective targets, allowing for membrane transfer [71–73]. Tunneling nanotubes are observed in several cell types, indicating that they can provide a general mode of intercellular communication. However, further research on the phenomena of intercellular communication is still needed.

4.4 Membrane vesicles as a mode of transfer

These pieces of the membrane have been reported to travel from cell to cell. Proteins and lipids destined for transfer are concentrated on the plasma membrane, resulting in the formation of an extracellular vesicle, which travels to distant cells (**Figure 6**).

This mode of melanin transfer is usually not considered a mode of pigment transfer. However, two studies suggest that melanosome-containing membrane vesicles are phagocytosed by keratinocytes or fused with the keratinocyte plasma membrane as a model of melanin transfer. Flow cytometry analysis of a human

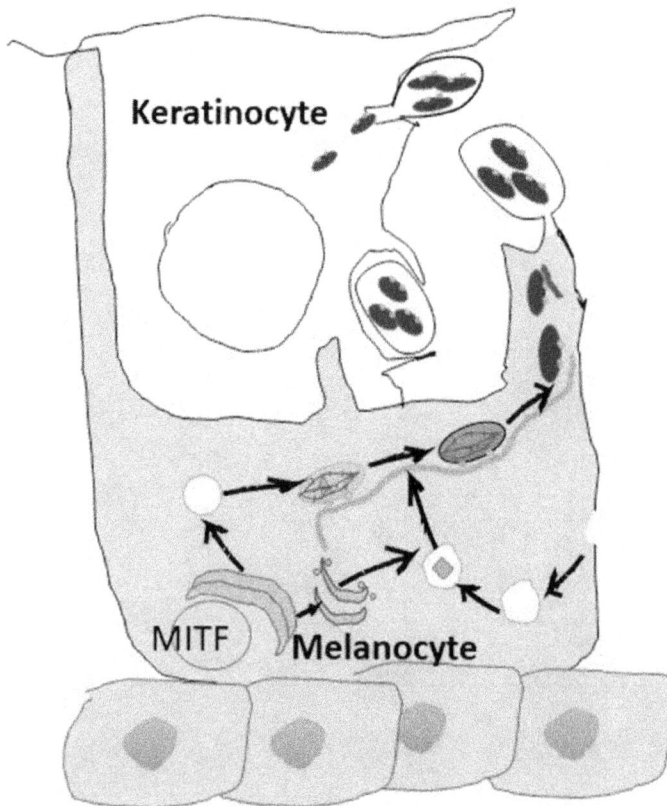

Figure 6.
Membrane vesicles as a mode of melanosomal transfer to the keratinocyte. Melanosomal vesicles either fuse with the plasma membrane of keratinocytes or ingest through phagocytosis.

melanoma cell line revealed that vesicles can be identified according to their size and fluorescent properties upon neoglycoprotein binding [74]. The addition of neo-glycoproteins could partially inhibit vesicle adhesion to keratinocytes because of the participation of carbohydrates in this interaction. The vesicles are finally swallowed by keratinocytes, thereby delivering melanin.

Another study showed that melanin transfers melanophores to fibroblasts [75]. Double membrane-covered melanin is transferred to distinct groups of recipient cells, some of which are located at a distance from the melanophore, suggesting that melanophores release melanin by the shedding of vesicles, and are subsequently recognized by fibroblasts through specific interactions.

Microparticles or microvesicles shed by live cells are believed to be formed upon the induction of cell stress, including cell activation and apoptosis, indicating true vectors of information exchange between cells [76]. This ubiquitous mode of material transfer for the delivery of melanosomes should be considered as a potential model.

In short, it is difficult to draw conclusions because none of the hypotheses provide concrete evidence. Naturally, these mechanisms are not mutually exclusive. Within this context, phagocytosis seems to be a necessary step for all proposed mechanisms, except for the fusion of plasma membranes.

5. Translocation, distribution, and degradation of melanosomes by the keratinocyte

After being transferred into recipient keratinocytes, melanosomes are selectively and predominantly translocated to the apical pole of the keratinocyte. As a result, they protect the underlying nucleus from mutagenic damage by absorbing UV light. This trafficking is mediated by cytoskeletal elements and microtubule-associated motor proteins. Studies have reported that dynein co-localizes with phagocytosed melanosomal aggregates throughout the cytoplasm, predominantly at the microtubule-organizing center in keratinocytes [24].

The distribution of recipient melanosomes within the keratinocytes varies according to complexion coloration, as demonstrated over a quarter of a century ago [77, 78]. Melanosomes are maintained as individual organelles throughout the cytosol of keratinocytes in the dark. In light-skinned individuals, melanosomes are significantly smaller and aggregated into membrane-bound clusters of 4–8 organelles. Whether these distinct distribution patterns are determined by factors within the transferred melanosome or are innate to the recipient keratinocytes remains unclear.

A recent study showed that the distribution pattern of recipient melanosomes is dictated by the type of donor melanocyte. An in vitro skin reconstruction model was assessed using combinations of keratinocytes and melanocytes from different complexion colorations [79]. In contrast, the skin type from which the recipient keratinocyte was derived regulates the distribution pattern of transferred mela-nosomes regardless of their size, as illustrated in the melanocyte/keratinocyte co-culture experiment [34]. However, the experimental context of these approaches casts doubt on the results. Therefore, the mechanism underlying the regulation of the distribution of melanosomes in the skin remains to be fully elucidated.

Melanosomes undergo degradation by the time they differentiate into corneocytes. As observed in the interface between the stratum granulosum and the stratum corneum, few melanosome structures remain in corneocytes of very darkly pigmented skin 49, and no apparent melanosomes remain in the corneocytes of light skin. There is a need to identify the hydrolytic processes used by keratinocytes to degrade the dense melanosome/melanin. Hydrolytic enzymes have been found to be involved in melanosome degradation by keratinocytes [80].

6. Dermatological evidence of abnormal pigmentation due to abnormalities in keratinocytes

6.1 EBS with mottled pigmentation (EBS-MP)

Epidermolysis bullosa simplex (EBS) is an autosomal dominant inherited skin disease characterized by blistering. EBS with mottled pigmentation (EBS-MP) is a rare form of generalized EBS characterized by non-scarring blistering and small hyper- and hypopigmented spots that form a mottled to the reticulate pattern. Using electron microscopy, an increased number of melanosomes within basal keratinocytes, dermal macrophages, and Schwann cells has been reported in EBS-MP patients. EBS-MP mostly involves the distal extremities and progressive mottled hyperpigmentation. During the neonatal period, differentiating EBS-MP from other subtypes is challenging due to the fact that pigmentary changes usually start later in infancy or childhood [81]. In the literature, most cases of EBS-MP have been attributed to a heterozygous missense mutation of KRT5 that results in the substitution of proline 25 with leucine (p25L) in the nonhelical V1 domain of KRT5 [81]. A 25-year-old Japanese woman and her cousin and a Chinese family have been reported as rare cases of EBS-MP due to a 1649delG mutation in the KRT5 tail domain (V2) [82, 83]. **Figure** 7 shows a familial case of a 1-year-old girl and her father with EBS-MP, who developed mottled to reticulate pattern hyperpigmentation on the extremities, trunk, and face (**Figure** 7). Electron microscopy was used to evaluate a skin sample from the father (unpublished).

6.2 Dowling-Degos disease (DDD)

DDD is characterized by progressive and disfiguring hyperpigmentation primarily in the flexural areas, which is a rare autosomal dominant genodermatosis with variable penetrance due to haploinsufficiency of the KRT5 gene [84, 85] on chromosome 12q [84]. KRT5/KRT14 is a crucial element of the basal keratinocyte cytoskeleton. KRT5 dysfunction alters organelle transportation and epidermal differentiation. DDD usually develops after puberty and is clinically distinguished by brownish hyperpigmented macules that alter a reticular pattern. These macules are mostly located in the skin flexures (sub-mammary, axillae, and groin), cervical region, trunk, and anterior surface of the thighs and upper arms. Pinpoint papules with keratin plugs that simulate comedones are also found in the palmar, axillary, cervical, perioral, and gluteal regions. Some studies have shown that DDD is associated with Hidradenitis suppurativa and can develop depressed perioral scars [84, 86, 87]. An interconnected hyperpigmented epidermal proliferation projection in the dermis denoted as the "antler-like" pattern, is characterized by a filiform pattern and is a key differentiating feature. In the case of DDD, hyperpigmented

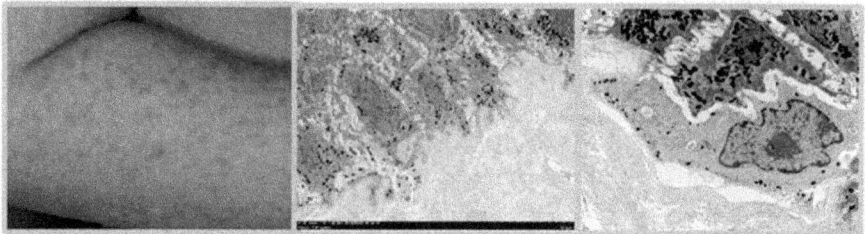

Figure 7.
Many brown macules scattered on arms, legs and face of a 1-year-old girl with EBS-MP.

proliferation is derived from both the epidermis and the follicular wall. This feature makes DDD different from other keratinized disorders. At present, there have not been any successful medicinal therapeutic approaches reported in the literature.

6.3 Galli-Galli disease (GGD)

GGD is a rare autosomal dominant genodermatosis that is considered an acantholytic variant of DDD, characterized by hyperkeratotic papules and progressive reticular hyperpigmentation involving the neck, trunk, and proximal extremities. GGD is associated with autosomal dominant mutations in KRT5, POGLUT1, and POFUT1 genes, and is clinically characterized as red-to-brown hyperkeratotic papules that gradually develop into brown reticulate lentigo-like macules over the trunk, neck, flexor, and extensor surfaces of the extremities. Severe flares and more diffuse distribution of cutaneous participation may be observed in rare cases [88]. A successful medicinal therapeutic approach has yet to be reported in the literature.

7. Potential mechanism of keratin mutation affects melanin distribution

It has been reported that the KRT5 head domain is noticeably more stable than the KRT14 head, and its distribution is altered following the depolymerization of microtubules. Several studies have reported that the KRT5 head domain interacts with heat shock cognate 70 (HSC70) and is involved in organelle transport [8, 89] and chaperone-mediated autophagy. In this context, the KRT5 mutation may affect melanosome degradation by modulating the interaction between KRT5 and HSC70, resulting in an abnormal accumulation of melanin in the keratinocytes (**Figure 8**).

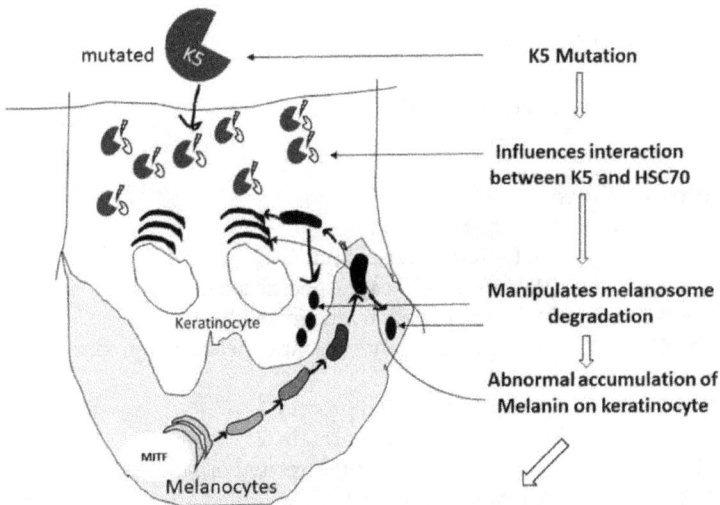

Figure 8.
Speculated mechanism of how keratin mutation affects melanin distribution.

8. Conclusion

The regulation of pigmentation in melanogenesis in melanocytes has been thoroughly investigated. However, the mechanism by which melanosomes are transferred into keratinocytes and the interaction among different molecules during this transfer process have yet to be well characterized. Recently, keratinocytes have become an interesting subject for pigmentary disorders. Surprisingly, little is known about the mechanism by which human melanosomes are transferred to keratinocytes and the degradation of melanosomes inside keratinocytes, and how genetic abnormalities in keratinocytes, but not in melanocytes, cause pigmentary skin disorders. This chapter covers the most important aspects of melanocytes and keratinocytes that induce pigmentation and summarizes the mechanism underlying the transfer of melanosomes to keratinocytes and the degradation of melanosomes inside keratinocytes. How genetic abnormalities in keratinocytes affect pigmentary skin disorders are also discussed in an attempt to shed light on hereditary pigmentary disorders and provide a conceptual framework for the role of keratinocytes in pigmentary disorders.

Acknowledgements

I would like to express my gratitude to my primary supervisor, Professor Mayumi Komine, and my Chairman, Professor Mamitaro Ohtsuki, who guided me throughout this work. I would also like to express my deep appreciation to my family and colleagues who helped me finalize my book chapter.

Funding

This research did not receive any external funding.

Conflicts of interest

The author declares no conflict of interest.

Abbreviations

ER	Endoplasmic reticulum
PMEL17	Melanocytes lineage-specific antigen gp100
MART-1	Melanoma antigen recognized by T cell-1
GPNMB	Glycoprotein nonmetastatic melanoma protein b
TYR	Tyrosinase
TYRP1	Tyrosinase-related protein-1
TYRP2	Tyrosinase-related protein-2
BLOCs	Biogenesis of lysosome-related organelles complexes
AP-1	Activator protein 1
MLPH	Melanophilin
MYO5A	Myosin-Va
RAB27A	Ras-related protein Rab27A
MITF	Microphthalmia-associated transcription factor
MIICs	MHC class II compartments

Hsc70	Heat shock cognate 70
NO	Nitric oxide
CREB	cAMP response element-binding protein
MC1R	Melanocortin-1 receptor
PAX3	Paired box protein 3
SOX9, SOX10	Sex-determining region Y-box 9 and 10
LEF-1	Lymphoid enhancer-binding factor 1
WS2	Waardenburg syndrome type 2
Foxn1	Forkhead box protein N1
bFGF	Fibroblast growth factor
HGF	Hepatocyte growth factor
NGF	Nerve growth factor
SCF	Stem cell factor
LIF	Leukemia inhibitory factor
KGFR	Keratinocyte growth factor receptor
PAR-2	Protease (proteinase)-activated receptor-2
DKK1	Dickkopf 1
UV	Ultraviolet
NFR	Tumor necrosis factor receptor
EGFR	Epidermal growth factor receptor
PGE2	Prostaglandin E2
PGF2α	Prostaglandin F2α
GM-CSF	Granulocyte-macrophage colony stimulating factor
α-MSH	α-Melanocyte-stimulating hormone
ACTH	Adrenocorticotropic hormone
SNAP	Soluble NSF attachment proteins
VAMP2	Vesicle-associated membrane protein 2
cAMP	Cyclic adenosine monophosphate
KRT	Keratinocyte
POGLUT1	Protein O-Glucosyltransferase 1
POFUT1	Protein O-Fucosyltransferase 1

Author details

Md. Razib Hossain*, Miho Kimura-Sashikawa and Mayumi Komine
Department of Dermatology, Jichi Medical University, Tochigi, Japan

*Address all correspondence to: razib@jichi.ac.jp

IntechOpen

References

[1] Lim HW, Collins SAB, Resneck JS Jr, Bolognia JL, Hodge JA, Rohrer TA, et al. The burden of skin disease in the United States. Journal of the American Academy of Dermatology. 2017;**76**:958-972. DOI: 10.1016/j.jaad.2016.12.043

[2] Scott G, Jacobs S, Leopardi S, Anthony FA, Learn D, Malaviya R, et al. Effects of PGF2alpha on human melanocytes and regulation of the FP receptor by ultraviolet radiation. Experimental Cell Research. 2005;**304**:407-416. DOI: 10.1016/j.yexcr.2004.11.016

[3] Park HY, Kosmadaki M, Yaar M, Gilchrest BA. Cellular mechanisms regulating human melanogenesis. Cellular and Molecular Life Sciences. 2009;**66**:1493-1506. DOI: 10.1007/s00018-009-8703-8

[4] Fistarol SK, Itin PH. Disorders of pigmentation. Journal der Deutschen Dermatologischen Gesellschaft. 2010;**8**:187-201. DOI: 10.1111/j.1610-0387.2009.07137.x

[5] Chamcheu JC, Siddiqui IA, Syed DN, Adhami VM, Liovic M, Mukhtar H. Keratin gene mutations in disorders of human skin and its appendages. Archives of Biochemistry and Biophysics. 2011;**508**:123-137. DOI: 10.1016/j.abb.2010.12.019

[6] Moriggi M, Pastorelli L, Torretta E, Tontini GE, Capitanio D, Bogetto SF, et al. Contribution of extracellular matrix and signal mechanotransduction to epithelial cell damage in inflammatory bowel disease patients: A Proteomic Study. Proteomics. 2017;**17**:23-24. DOI: 10.1002/pmic.201700164

[7] Mashukova A, Forteza R, Salas PJ. Functional analysis of keratin-associated proteins in intestinal epithelia: Heat-shock protein chaperoning and kinase rescue. Methods in Enzymology. 2016;**569**:139-154. DOI: 10.1016/bs.mie.2015.08.019

[8] Planko L, Böhse K, Höhfeld J, Betz RC, Hanneken S, Eigelshoven S, et al. Identification of a keratin-associated protein with a putative role in vesicle transport. European Journal of Cell Biology. 2007;**86**:827-839. DOI: 10.1016/j.ejcb.2007.02.004

[9] Raposo G, Marks MS. Melanosomes--dark organelles enlighten endosomal membrane transport. Nature Reviews. Molecular Cell Biology. 2007;**8**:786-797. DOI: 10.1038/nrm2258

[10] Watt B, Tenza D, Lemmon MA, Kerje S, Raposo G, Andersson L, et al. Mutations in or near the transmembrane domain alter PMEL amyloid formation from functional to pathogenic. PLoS Genetics. 2011;7:e1002286. DOI: 10.1371/journal.pgen.1002286

[11] Bennett DC, Lamoreux ML. The color loci of mice: A genetic century. Pigment Cell Research. 2003;**16**:333-344. DOI: 10.1034/j.1600-0749.2003.00067.x

[12] Hearing VJ. Biogenesis of pigment granules: A sensitive way to regulate melanocyte function. Journal of Dermatological Science. 2005;**37**:3-14. DOI: 10.1016/j.jdermsci.2004.08.014

[13] Yamaguchi Y, Hearing VJ. Physiological factors that regulate skin pigmentation. BioFactors. 2009;**35**:193-199. DOI: 10.1002/biof.29

[14] Hoashi T, Watabe H, Muller J, Yamaguchi Y, Vieira WD, Hearing VJ. MART-1 is required for the function of the melanosomal matrix protein PMEL17/GP100 and the maturation of melanosomes. The Journal of Biological Chemistry. 2005;**280**:14006-14016. DOI: 10.1074/jbc.M413692200

[15] Hoashi T, Sato S, Yamaguchi Y, Passeron T, Tamaki K, Hearing VJ.

Glycoprotein nonmetastatic melanoma protein b, a melanocytic cell marker, is a melanosome-specific and proteolytically released protein. The FASEB Journal. 2010;24:1616-1629. DOI: 10.1096/fj.09-151019

[16] Zhang P, Liu W, Zhu C, Yuan X, Li D, Gu W, et al. Silencing of GPNMB by siRNA inhibits the formation of melanosomes in melanocytes in a MITF-independent fashion. PLoS One. 2012;7:e42955. DOI: 10.1371/journal.pone.0042955

[17] Chi A, Valencia JC, Hu ZZ, Watabe H, Yamaguchi H, Mangini NJ, et al. Proteomic and bioinformatic characterization of the biogenesis and function of melanosomes. Journal of Proteome Research. 2006;5:3135-3144. DOI: 10.1021/pr060363j

[18] Basrur V, Yang F, Kushimoto T, Higashimoto Y, Yasumoto K, Valencia J, et al. Proteomic analysis of early melanosomes: Identification of novel melanosomal proteins. Journal of Proteome Research. 2003;2:69-79. DOI: 10.1021/pr025562r

[19] Kushimoto T, Basrur V, Valencia J, Matsunaga J, Vieira WD, Ferrans VJ, et al. A model for melanosome biogenesis based on the purification and analysis of early melanosomes. Proceedings of the National Academy of Sciences of the United States of America. 2001;98:10698-10703. DOI: 10.1073/pnas.191184798

[20] Slominski A, Tobin DJ, Shibahara S, Wortsman J. Melanin pigmentation in mammalian skin and its hormonal regulation. Physiological Reviews. 2004;84:1155-1228. DOI: 10.1152/physrev.00044.2003

[21] Schallreuter KU, Kothari S, Chavan B, Spencer JD. Regulation of melanogenesis--controversies and new concepts. Experimental Dermatology. 2008;17:395-404. DOI: 10.1111/j.1600-0625.2007.00675.x

[22] Park HY, Perez JM, Laursen R, Hara M, Gilchrest BA. Protein kinase C-beta activates tyrosinase by phosphorylating serine residues in its cytoplasmic domain. The Journal of Biological Chemistry. 1999;274:16470-16478. DOI: 10.1074/jbc.274.23.16470

[23] Hara M, Yaar M, Byers HR, Goukassian D, Fine RE, Gonsalves J, et al. Kinesin participates in melanosomal movement along melanocyte dendrites. The Journal of Investigative Dermatology. 2000;114:438-443. DOI: 10.1046/j.1523-1747.2000.00894.x

[24] Byers HR, Yaar M, Eller MS, Jalbert NL, Gilchrest BA. Role of cytoplasmic dynein in melanosome transport in human melanocytes. The Journal of Investigative Dermatology. 2000;114:990-997. DOI: 10.1046/j.1523-1747.2000.00957.x

[25] Barral DC, Seabra MC. The melanosome as a model to study organelle motility in mammals. Pigment Cell Research. 2004;17:111-118. DOI: 10.1111/j.1600-0749.2004.00138.x

[26] Van Den Bossche K, Naeyaert JM, Lambert J. The quest for the mechanism of melanin transfer. Traffic. 2006;7:769-778. DOI: 10.1111/j.1600-0854.2006.00425.x

[27] Kuroda TS, Fukuda M. Rab27A-binding protein Slp2-a is required for peripheral melanosome distribution and elongated cell shape in melanocytes. Nature Cell Biology. 2004;6:1195-1203. DOI: 10.1038/ncb1197

[28] Kuroda TS, Ariga H, Fukuda M. The actin-binding domain of Slac2-a/melanophilin is required for melanosome distribution in melanocytes. Molecular and Cellular Biology. 2003;23:5245-5255. DOI: 10.1128/mcb.23.15.5245-5255.2003

[29] Kondo T, Hearing VJ. Update on the regulation of mammalian melanocyte

function and skin pigmentation. Expert Review of Dermatology. 2011;**6**:97-108. DOI: 10.1586/edm.10.70

[30] Vachtenheim J, Borovanský J. "Transcription physiology" of pigment formation in melanocytes: Central role of MITF. Experimental Dermatology. 2010;**19**:617-627. DOI: 10.1111/j.1600-0625.2009.01053.x

[31] Steingrímsson E, Copeland NG, Jenkins NA. Melanocytes and the microphthalmia transcription factor network. Annual Review of Genetics. 2004;**38**:365-411. DOI: 10.1146/annurev.genet.38.072902.092717

[32] Rouzaud F, Costin GE, Yamaguchi Y, Valencia JC, Berens WF, Chen KG, et al. Regulation of constitutive and UVR-induced skin pigmentation by melanocortin 1 receptor isoforms. The FASEB Journal. 2006;**20**:1927-1929. DOI: 10.1096/fj.06-5922fje

[33] Yamaguchi Y, Hearing VJ, Itami S, Yoshikawa K, Katayama I. Mesenchymal-epithelial interactions in the skin: Aiming for site-specific tissue regeneration. Journal of Dermatological Science. 2005;**40**:1-9. DOI: 10.1016/j.jdermsci.2005.04.006

[34] Minwalla L, Zhao Y, Le Poole IC, Wickett RR, Boissy RE. Keratinocytes play a role in regulating distribution patterns of recipient melanosomes in vitro. The Journal of Investigative Dermatology. 2001;**117**:341-347. DOI: 10.1046/j.0022-202x.2001.01411.x

[35] Yoshida Y, Hachiya A, Sriwiriyanont P, Ohuchi A, Kitahara T, Takema Y, et al. Functional analysis of keratinocytes in skin color using a human skin substitute model composed of cells derived from different skin pigmentation types. The FASEB Journal. 2007;**21**:2829-2839. DOI: 10.1096/fj.06-6845com

[36] Weiner L, Han R, Scicchitano BM, Li J, Hasegawa K, Grossi M, et al. Dedicated epithelial recipient cells determine pigmentation patterns. Cell. 2007;**130**:932-942. DOI: 10.1016/j.cell.2007.07.024

[37] Hirobe T. Role of keratinocyte-derived factors involved in regulating the proliferation and differentiation of mammalian epidermal melanocytes. Pigment Cell Research. 2005;**18**:2-12. DOI: 10.1111/j.1600-0749.2004.00198.x

[38] Scott G, Leopardi S, Parker L, Babiarz L, Seiberg M, Han R. The proteinase-activated receptor-2 mediates phagocytosis in a Rho-dependent manner in human keratinocytes. The Journal of Investigative Dermatology. 2003;**121**:529-541. DOI: 10.1046/j.1523-1747.2003.12427.x

[39] Cardinali G, Ceccarelli S, Kovacs D, Aspite N, Lotti LV, Torrisi MR, et al. Keratinocyte growth factor promotes melanosome transfer to keratinocytes. The Journal of Investigative Dermatology. 2005;**125**:1190-1199. DOI: 10.1111/j.0022-202X.2005.23929.x

[40] Scott G, Leopardi S, Printup S, Malhi N, Seiberg M, Lapoint R. Proteinase-activated receptor-2 stimulates prostaglandin production in keratinocytes: Analysis of prostaglandin receptors on human melanocytes and effects of PGE2 and PGF2alpha on melanocyte dendricity. The Journal of Investigative Dermatology. 2004;**122**:1214-1224. DOI: 10.1111/j.0022-202X.2004.22516.x

[41] Glinka A, Wu W, Delius H, Monaghan AP, Blumenstock C, Niehrs C. Dickkopf-1 is a member of a new family of secreted proteins and functions in head induction. Nature. 1998;**391**:357-362. DOI: 10.1038/34848

[42] Burgoyne RD, Morgan A. Secretory granule exocytosis. Physiological Reviews. 2003;**83**:581-632. DOI: 10.1152/physrev.00031.2002

[43] Yamamoto O, Bhawan J. Three modes of melanosome transfers in Caucasian facial skin: Hypothesis based on an ultrastructural study. Pigment Cell Research. 1994;7:158-169. DOI: 10.1111/j.1600-0749.1994.tb00044.x

[44] Swift JA. Transfer of melanin granules from melanocytes to the cortical cells of human hair. Nature. 1964;203:976-977. DOI: 10.1038/203976b0

[45] Virador VM, Muller J, Wu X, Abdel-Malek ZA, Yu ZX, Ferrans VJ, et al. Influence of alpha-melanocyte-stimulating hormone and ultraviolet radiation on the transfer of melanosomes to keratinocytes. The FASEB Journal. 2002;16:105-107. DOI: 10.1096/fj.01-0518fje

[46] Quast T, Wehner S, Kirfel G, Jaeger K, De Luca M, Herzog V. sAPP as a regulator of dendrite motility and melanin release in epidermal melanocytes and melanoma cells. The FASEB Journal. 2003;17:1739-1741. DOI: 10.1096/fj.02-1059fje

[47] Scott G, Zhao Q. Rab3a and SNARE proteins: Potential regulators of melanosome movement. The Journal of Investigative Dermatology. 2001;116:296-304. DOI: 10.1046/j.1523-1747.2001.01221.x

[48] Wade N, Bryant NJ, Connolly LM, Simpson RJ, Luzio JP, Piper RC, et al. Syntaxin 7 complexes with mouse Vps10p tail interactor 1b, syntaxin 6, vesicle-associated membrane protein (VAMP)8, and VAMP7 in b16 melanoma cells. The Journal of Biological Chemistry. 2001;276:19820-19827. DOI: 10.1074/jbc.M010838200

[49] Tarafder AK, Bolasco G, Correia MS, Pereira FJC, Iannone L, Hume AN, et al. Rab11b mediates melanin transfer between donor melanocytes and acceptor keratinocytes via coupled exo/endocytosis. The

Journal of Investigative Dermatology. 2014;134:1056-1066. DOI: 10.1038/jid.2013.432

[50] Wu B, Guo W. The exocyst at a glance. Journal of Cell Science. 2015;128:2957-2964. DOI: 10.1242/jcs.156398

[51] Moreiras H, Pereira FJC, Neto MV, Bento-Lopes L, Festas TC, Seabra MC, et al. The exocyst is required for melanin exocytosis from melanocytes and transfer to keratinocytes. Pigment Cell & Melanoma Research. 2020;33:366-371. DOI: 10.1111/pcmr.12840

[52] Schlüter OM, Khvotchev M, Jahn R, Südhof TC. Localization versus function of Rab3 proteins. Evidence for a common regulatory role in controlling fusion. The Journal of Biological Chemistry. 2002;277:40919-40929. DOI: 10.1074/jbc.M203704200

[53] Martelli AM, Baldini G, Tabellini G, Koticha D, Bareggi R, Baldini G. Rab3A and Rab3D control the total granule number and the fraction of granules docked at the plasma membrane in PC12 cells. Traffic. 2000;1:976-986

[54] Araki K, Horikawa T, Chakraborty AK, Nakagawa K, Itoh H, Oka M, et al. Small Gtpase rab3A is associated with melanosomes in melanoma cells. Pigment Cell Research. 2000;13:332-336. DOI: 10.1034/j.1600-0749.2000.130505.x

[55] Blois MS. Phagocytosis of melanin particles by human epidermal cells in vitro. The Journal of Investigative Dermatology. 1968;50:336-337. DOI: 10.1038/jid.1968.54

[56] Potter B, Medenica M. Ultramicroscopic phagocytosis of synthetic melanin by epidermal cells in vivo. The Journal of Investigative Dermatology. 1968;51:300-303. DOI: 10.1038/jid.1968.132

[57] Wolff K, Jimbow K, Fitzpatrick TB. Experimental pigment donation in vivo.

Journal of Ultrastructure Research. 1974;**47**:400-419. DOI: 10.1016/s0022-5320(74)90017-3

[58] Wolff K, Konrad K. Melanin pigmentation: An in vivo model for studies of melanosome kinetics within keratinocytes. Science. 1971;**174**:1034-1035. DOI: 10.1126/science.174.4013.1034

[59] Blott EJ, Griffiths GM. Secretory lysosomes. Nature Reviews. Molecular Cell Biology. 2002;**3**:122-131. DOI: 10.1038/nrm732

[60] Rabinovitch M. Professional and non-professional phagocytes: An introduction. Trends in Cell Biology. 1995;**5**:85-87. DOI: 10.1016/s0962-8924(00)88955-2

[61] Okazaki K, Uzuka M, Morikawa F, Toda K, Seiji M. Transfer mechanism of melanosomes in epidermal cell culture. The Journal of Investigative Dermatology. 1976;**67**:541-547. DOI: 10.1111/1523-1747.ep12664554

[62] Birbeck MS, Mercer EH, Barnicot NA. The structure and formation of pigment granules in human hair. Experimental Cell Research. 1956;**10**:505-514. DOI: 10.1016/0014-4827(56)90022-2

[63] Ruprecht KW. Pigmentation of the down feather in Gallus domesticus L. Light and electron microscopic studies of melanosome transfer. Zeitschrift für Zellforschung und Mikroskopische Anatomie. 1971;**112**:396-413

[64] Mottaz JH, Zelickson AS. Melanin transfer: A possible phagocytic process. The Journal of Investigative Dermatology. 1967;**49**:605-610. DOI: 10.1038/jid.1967.187

[65] Seiberg M, Paine C, Sharlow E, Andrade-Gordon P, Costanzo M, Eisinger M, et al. The protease-activated receptor 2 regulates pigmentation via keratinocyte-melanocyte interactions. Experimental Cell Research. 2000;**254**: 25-32. DOI: 10.1006/excr.1999.4692

[66] Boissy RE. Melanosome transfer to and translocation in the keratinocyte. Experimental Dermatology. 2003;**12**(Suppl. 2):5-12. DOI: 10.1034/j.1600-0625.12.s2.1.x

[67] Ando H, Niki Y, Yoshida M, Ito M, Akiyama K, Kim JH, et al. Keratinocytes in culture accumulate phagocytosed melanosomes in the perinuclear area. Pigment Cell & Melanoma Research. 2010;**23**:129-133. DOI: 10.1111/j.1755-148X.2009.00640.x

[68] Bhawan J. Ultrastructure of melanocyte-keratinocyte interactions in pigmented basal-cell carcinoma. Proceedings of Yale Journal of Biology and Medicine:548

[69] Scott G, Leopardi S, Printup S, Madden BC. Filopodia are conduits for melanosome transfer to keratinocytes. Journal of Cell Science. 2002;**115**: 1441-1451

[70] Rustom A, Saffrich R, Markovic I, Walther P, Gerdes HH. Nanotubular highways for intercellular organelle transport. Science. 2004;**303**:1007-1010. DOI: 10.1126/science.1093133

[71] Stinchcombe JC, Bossi G, Booth S, Griffiths GM. The immunological synapse of CTL contains a secretory domain and membrane bridges. Immunity. 2001;**15**:751-761. DOI: 10.1016/s1074-7613(01)00234-5

[72] Bossi G, Trambas C, Booth S, Clark R, Stinchcombe J, Griffiths GM. The secretory synapse: The secrets of a serial killer. Immunological Reviews. 2002;**189**:152-160. DOI: 10.1034/j.1600-065x.2002.18913.x

[73] Onfelt B, Nedvetzki S, Yanagi K, Davis DM. Cutting edge: Membrane nanotubes connect immune cells.

Journal of Immunology. 2004;**173**:1511-1513. DOI: 10.4049/jimmunol.173.3.1511

[74] Cerdan D, Redziniak G, Bourgeois CA, Monsigny M, Kieda C. C32 human melanoma cell endogenous lectins: Characterization and implication in vesicle-mediated melanin transfer to keratinocytes. Experimental Cell Research. 1992;**203**:164-173. DOI: 10.1016/0014-4827(92)90052-a

[75] Aspengren S, Hedberg D, Wallin M. Studies of pigment transfer between Xenopus laevis melanophores and fibroblasts in vitro and in vivo. Pigment Cell Research. 2006;**19**:136-145. DOI: 10.1111/j.1600-0749.2005.00290.x

[76] Hugel B, Martínez MC, Kunzelmann C, Freyssinet JM. Membrane microparticles: Two sides of the coin. Physiology (Bethesda). 2005;**20**:22-27. DOI: 10.1152/physiol.00029.2004

[77] Szabó G, Gerald AB, Pathak MA, Fitzpatrick TB. Racial differences in the fate of melanosomes in human epidermis. Nature. 1969;**222**:1081-1082. DOI: 10.1038/2221081a0

[78] Konrad K, Wolff K. Hyperpigmentation, melanosome size, and distribution patterns of melanosomes. Archives of Dermatology. 1973;**107**:853-860

[79] Bessou-Touya S, Picardo M, Maresca V, Surlève-Bazeille JE, Pain C, Taïeb A. Chimeric human epidermal reconstructs to study the role of melanocytes and keratinocytes in pigmentation and photoprotection. The Journal of Investigative Dermatology. 1998;**111**:1103-1108. DOI: 10.1046/j.1523-1747.1998.00405.x

[80] Hori Y, Toda K, Pathak MA, Clark WH Jr, Fitzpatrick TB. A fine-structure study of the human epidermal melanosome complex and its acid phosphatase activity. Journal of Ultrastructure Research. 1968;**25**:109-120. DOI: 10.1016/s0022-5320(68)80064-4

[81] Echeverría-García B, Vicente A, Hernández Á, Mascaró JM, Colmenero I, Terrón A, et al. Epidermolysis bullosa simplex with mottled pigmentation: A family report and review. Pediatric Dermatology. 2013;**30**:e125-e131. DOI: 10.1111/j.1525-1470.2012.01748.x

[82] Horiguchi Y, Sawamura D, Mori R, Nakamura H, Takahashi K, Shimizu H. Clinical heterogeneity of 1649delG mutation in the tail domain of keratin 5: A Japanese family with epidermolysis bullosa simplex with mottled pigmentation. The Journal of Investigative Dermatology. 2005;**125**:83-85. DOI: 10.1111/j.0022-202X.2005.23790.x

[83] Tang HY, Du WD, Cui Y, Fan X, Quan C, Fang QY, et al. One novel and two recurrent mutations in the keratin 5 gene identified in Chinese patients with epidermolysis bullosa simplex. Clinical and Experimental Dermatology. 2009;**34**:e957-e961. DOI: 10.1111/j.1365-2230.2009.03703.x

[84] Betz RC, Planko L, Eigelshoven S, Hanneken S, Pasternack SM, Bussow H, et al. Loss-of-function mutations in the keratin 5 gene lead to Dowling-Degos disease. American Journal of Human Genetics. 2006;**78**:510-519. DOI: 10.1086/500850

[85] Liao H, Zhao Y, Baty DU, McGrath JA, Mellerio JE, McLean WH. A heterozygous frameshift mutation in the V1 domain of keratin 5 in a family with Dowling-Degos disease. The Journal of Investigative Dermatology. 2007;**127**:298-300. DOI: 10.1038/sj.jid.5700523

[86] Arjona-Aguilera C, Linares-Barrios M, Albarrán-Planelles C, Jiménez-Gallo D. Dowling-Degos

disease associated with hidradenitis suppurativa: A case report. Actas Dermo-Sifiliográficas. 2015;**106**:337-338. DOI: 10.1016/j.ad.2014.09.010

[87] Valdés F, Peteiro C, Toribio J. Enfermedad de Dowling-Degos. Actas Dermo-Sifiliográficas. 2003;**94**:409-411. DOI: 0.1016/S0001-7310(03)76713-1

[88] Rundle CW, Ophaug S, Simpson EL. Acitretin therapy for Galli-Galli disease. JAAD Case Rep. 2020;**6**:457-461. DOI: 10.1016/j.jdcr.2020.02.042

[89] Salas PJ, Forteza R, Mashukova A. Multiple roles for keratin intermediate filaments in the regulation of epithelial barrier function and apico-basal polarity. Tissue Barriers. 2016;**4**: e1178368. DOI: 10.1080/ 21688370.2016.1178368

Section 2

Epidermis and Inflammation

Chapter 3

Effects of IL-17 on Epidermal Development

Emi Sato and Shinichi Imafuku

Abstract

Immunotherapies targeting interleukin 17 (IL-17) have a strong effect on plaque psoriasis. However, many previous studies on IL-17 focused only on the T-helper 17 (Th17) immune response, and a few studies have reported that IL-17A may affect psoriatic epidermal structure. IL-17 includes six family members, namely IL-17A–F, which are involved in a wide variety of biological responses. IL-17A is produced mainly by Th17 cells or group 3 innate lymphoid cells (ILC3), while IL-17C is locally produced by epithelial cells, such as keratinocytes. In contrast to IL-17C, which is locally produced in various cells such as keratinocytes, it is predicted that IL-17A, which is produced by limited cells and has systemic effects, has different roles in epidermal development. For example, several research studies have shown that IL-17A affects terminal differentiation of epidermis by suppressing the expression of filaggrin or loricrin in keratinocytes. On the other hand, IL-17C, which is produced by keratinocytes themselves, does not have as strong as an effect on epidermal development as IL-17A. In this chapter, we summarized the effects of IL-17A and other IL-17 members on epidermal development and their comprehensive roles based on previously reported papers.

Keywords: psoriasis, IL-17, epidermal structure, terminal differentiation genes

1. Introduction

Since 2003, biologics targeting inflammatory cytokines have been used to treat a wide range of diseases, especially those in which excessive autoimmunity is involved in the pathogenesis, such as rheumatoid arthritis, inflammatory bowel disease, psoriasis, and atopic dermatitis [1–6]. Before biologics targeting specific pro-inflammatory cytokines were introduced, the mechanisms involved in the pathogenesis of diseases were understood to a certain extent. However, once the therapy was actually started, various contradictions were seen. For example, psoriasis is thought to be mainly caused by T-helper 1 (Th1) and T-helper 17 (Th17) cells [7–9]; therefore, treatments that inhibit the differentiation of undifferentiated T-cells into Th1 and Th17 (e.g., CD80/86, CD2, CD11a, and RORγT inhibitors) were expected to be the most effective in correcting the root of the disease. However, contrary to expectations, biologics targeting interleukin 17 (IL-17) are actually considered to be the most effective treatment for psoriasis [1]. In other words, direct inhibition of terminal IL-17 is more effective than inhibition of Th17 differentiation in psoriasis, as suggested by actual patient data. This suggested that IL-17, especially IL-17A, might be produced by cells other than Th17 cells. Subsequent studies also revealed that group 3 innate lymphoid cells (ILC3) and γδ T-cells in the skin produce IL-17A [10–12]. Although new facts may be revealed in the future, this chapter summarizes our current knowledge of the role of IL-17 and its sources of production, with a focus on epidermal development.

2. Which cells produce IL-17 in the skin?

IL-17 is a homodimeric glycoprotein consisting of 20–30 kDa peptides. The IL-17 family can be divided into six family members, IL-17A-IL-17F (**Table 1**). This section summarizes the roles of each IL-17 member and the cells that produce them (**Figure 1**).

2.1 IL-17A

Th17 cells are thought to be the major producers of IL-17A. However, the biologics that target IL-17A have a greater effect on psoriasis compared with subunits, such as p40, that are required for differentiation into Th17 [13]. This suggests that cells other than Th17 cells are also responsible for IL-17A production. Non-Th17 cells that have been reported in the past include: neutrophils [14], mast cells [15, 16], CD8+ T-cells [17], αβ T-cells [18], γδ T-cells [12], and innate lymphoid cells (ILC) [10, 11]. IL-17A has effects on various cells, such as fibroblasts, keratinocytes, endothelial cells, and macrophages, to induce release of inflammatory cytokines/chemokines such as IL-6, TNF-α, and IL-8, as well as promoting neutrophil migration to induce inflammation [19]. In particular, IL-17A induces granulocyte-colony stimulating factor (G-CSF) and IL-8, cytokines that are strongly involved in the activation of neutrophils [19–21]. The inflammation that is induced by IL-17A contributes greatly to the defense against bacterial, fungal, and viral infections, but excessive inflammation can lead to autoimmune diseases and the creation of unfortunate conditions in the skin, such as psoriasis and hidradenitis suppurativa (HS), as described below. Additionally, IL-17A is directly involved in epidermal differentiation, as described below, and this may also be related to the formation of a pathological epidermis, such as psoriasis [22–24].

Ligand	Receptor	Producer cells	Effects
IL-17A	IL-17RA/RC	T cells (RORγt+ (**Th17,Tc17**), γδ+, αβ+), innate lymphoid cells (**ILC3**), neutrophils, and mast cells	protection against pathogens and inflammation, epidermal proliferation, and differentiation
IL-17B	IL-17RB	tumor cells (lung cancer, breast cancer, and leukemia), intestinal epithelium, chondrocyte and neuron	embryonic development, tissue regeneration, inflammation, and tumorgenesis
IL-17C	IL-17RA/RE	keratinocytes, epithelial cells, dendritic cell, macrophages, and CD4+ T-cells	protection against pathogens and inflammation
IL-17D	Unknown	resting CD4+ T cells and resting CD19+ B cells	inhibition of hematopoiesis
IL-17E (IL-25)	IL-17RA/RB	T cells (GATA3+ (**Th2**), CD8+), mast cells, eosinophils, epithelial cells and endothelial cells	Th2 type immune response
IL-17F	IL-17RA/RC	T cells (Th17, Tc17) and innate lymphoid cells (ILC3)	protection against pathogens and inflammation

Bolded areas indicate important producing cells.

Table 1.
Overview of the Human IL-17 Family Members.

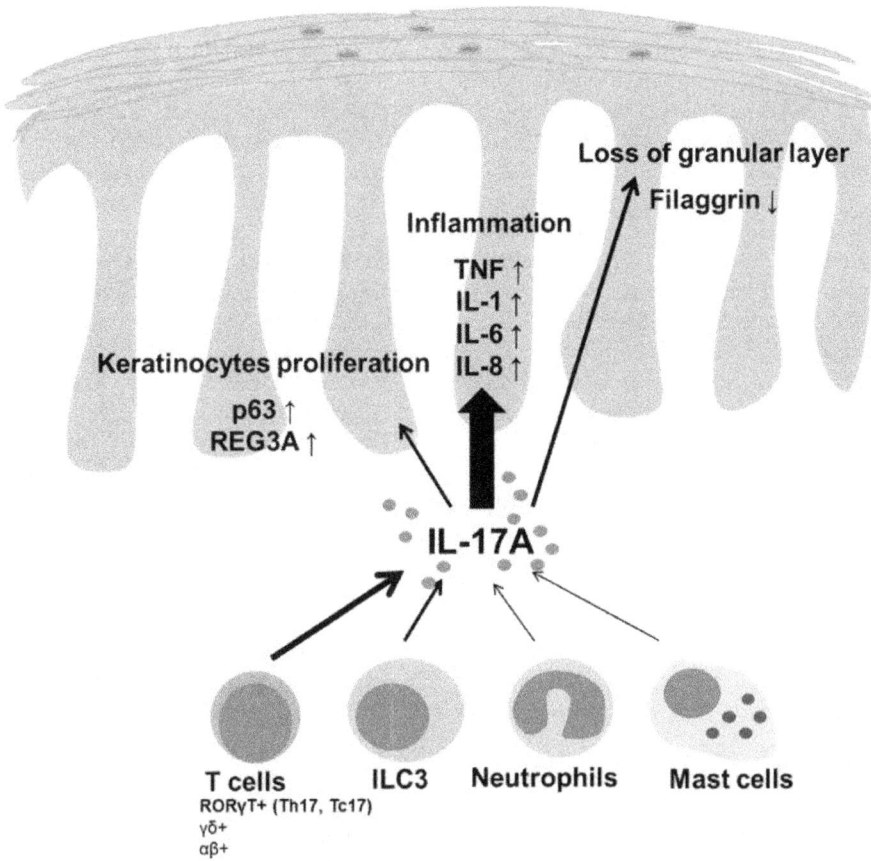

Figure 1.
The role of IL-17A in the skin. Immune cells, mainly Th17 cells, produce IL-17A, which promotes inflammation and proliferation of the epidermis, inhibits terminal differentiation, and causes loss of the granular layer.

2.2 IL-17B

Compared with IL-17A, IL-17B has not been studied as much, and so, its role remains unknown. IL-17B is the second most homologous to IL-17A, after IL-17F, but it is not produced by Th17 cells [25, 26]. In humans, IL-17B mRNA expression was confirmed in the tissues of the pancreas, small intestine, and stomach [27]. Although IL-17B did not induce a strong production of IL-6 directly from fibroblasts as IL-17A did, it did enhance the production of G-CSF and IL-6 in TNF-α-stimulated fibroblasts [27, 28]. However, there are reports that IL-17B has anti-inflammatory effects and is protective against asthma and colitis [29]. Recently, it was reported that IL-17B is positioned as a new component in the regulation of human type 2 immunity [30]. IL-17B has also been reported to promote the progression of malignant tumors such as lung cancer and breast cancer [31, 32]. The role of IL-17B on the epidermis is unknown and needs to be verified in the future.

2.3 IL-17C

IL-17C is known to be expressed by epithelial cells including keratinocytes, innate immune cells such as dendritic cells and macrophages, and CD4+ T-cells [19]. IL-17C stimulates Th17 T-cells to increase synthesis of IL-17A/F and IL-22,

and then IL-17A strongly induces IL-17C in keratinocytes, forming an inflammatory loop [33]. One study reported that overexpression of IL-17C in keratinocytes
promotes psoriasiform skin inflammation in mouse experiments [34]. It has
also been reported that neutralizing IL-17C suppressed atopic dermatitis and
psoriasis-like dermatitis in mice [35]. IL-17C neutralizing antibody, MOR106,
inhibits both Th2-type and Th17-type immune responses; and clinical trials for
atopic dermatitis are being conducted against the IL-17C neutralizing antibody
MOR106 [33, 36]. In a three-dimensional human epidermal model using normal
human keratinocytes, nine genes, including S100A7A, were reported to be commonly upregulated by IL-17A and IL-17C, but IL-17A induced the loss of the
granular layer in the epidermis, whereas IL-17C did not induce such epidermal
changes [22]. Also, although IL-17A is protective against fungal infections, IL-17C
deficiency is reported as lethal against systemic infection with Candida albicans
by IL-17C deficient mouse model [37]. However, another study using IL-17C
deficient mice reported that IL-17C is not involved in immunity against Candida
infection [38]. In murine bacterial infection models, IL-17C has been shown to be
protective against infection by Citrobacter rodentium, Pseudomonas aeruginosa,
and Haemophilus influenzae [39–41]. In viral infections, ex vivo administration of
IL-17C to mice reduced the apoptosis of neurons caused by HSV-2 infection, suggesting that keratinocyte-derived IL-17C acts to protect nerve fibers during HSV-2
reactivation in the skin [42]. It is also known that IL-17C is secreted through
NF-κB activation by stimulation of pathogen-associated molecular patterns
(PAMPs) [43] and is secreted in an autocrine manner to play a role in the initial
defense response against pathogens [44, 45].

2.4 IL-17D

IL-17D has not been studied as much as IL-17A [46]. IL-17D was found to be
highly expressed in the skeletal muscle, brain, adipose tissue, heart, lung, and
pancreas, while low in the bone marrow, fetal liver, kidney, leukocyte, liver, lymph
node, placenta, spleen, thymus, and tonsillar tissue [47]. IL-17D is also expressed
at low levels on resting CD4+ T-cells and resting CD19+ B cells. IL-17D was poorly
expressed on activated CD4+ T-cells, resting and activated CD8+ T-cells, resting
and activated CD14+ monocytes, and activated CD19+ B cells. [47]. IL-17D, like
other IL-17 family members, was capable of stimulating the production of other
cytokines such as IL-6, IL-8, and granulocyte-macrophage colony-stimulating
factor (GM-CSF) [48]. IL-17D stimulates human umbilical vein endothelial cells
to produce IL-8 to levels in the physiological range that inhibit hematopoiesis [49].
However, it has also been reported that the level of GM-CSF produced by human
umbilical vein endothelial cells is much lower than that required to stimulate
myeloid proliferation [50]. The expression of IL-17D mRNA has been found to be
decreased in psoriatic skin, but the reason for this is still unresolved [51].

2.5 IL-17E

IL-17E, also known as IL-25, has the lowest homology with IL-17A, among its
other five family members [25, 26]. The most important difference between IL-17E
and IL-17A is that IL-17E is involved in the Th2 type immune response [19]. On
the other hand, IL-17E induces immune responses by IL-5 and IL-13 through ILC2
in a T cell-independent manner [52]. In autoimmune diseases, IL-17E has been
thought to negatively regulate pathogenesis by suppressing Th17-type immune
responses [53]. However, it was recently reported that IL-17E is highly expressed in
the epidermis of imiquimod-induced psoriasis in mice, and that IL-17E promotes

keratinocyte proliferation and inflammatory responses, leading to psoriasis-like skin inflammation [54]. These two contradictory reports are both based on experiments using IL-17E deficient mice, but need to be validated in actual human psoriasis epidermis in the future.

2.6 IL-17F

IL-17F has the highest homology (40–55%) with IL-17A, followed by IL-17B (29%), IL-17D (25%), IL-17C (23%), and IL-17E (17%) [25, 26]. Like IL-17A, IL-17F binds to the IL-17RA/IL-17RC heterodimer receptor. Also, studies have known that IL-17F forms a homodimer or heterodimer with IL-17A when it binds to the receptor [55, 56]. As well as loss-of-function mutations in IL-17RA, it has been confirmed that partial loss-of-function mutations in IL-17F cause chronic mucocutaneous candidiasis [57]. IL-17F induces a variety of inflammatory molecules such as IL-6 and IL-8 via NF-κB, mitogen-activated protein kinase (MAPK), and CCAAT/ enhancer binding protein (C/EBP) [19]. IL-17F is derived primarily from Th17, Tc17, and ILC3 to enhance skin inflammation [58]. Bimekizumab (a biologic of both IL-17A and IL-17F) has been reported to be more effective than blocking IL-17A or IL-17F alone, particularly in inhibiting neutrophil chemotaxis and activation of synovial cells or human dermal fibroblasts in vitro [59].

3. What effect does IL-17 have on the epidermis?

3.1 Inflammation

IL-17A and IL-17F bind to IL-17RA/IL-17RC heterodimeric receptors and activate NF-κB, MAPK, and C/EBP, thereby inducing inflammation with cytokines such as IL-1, IL-6, and TNF-α and chemokines such as IL-8 (CXCL-8) and CXCL-1. These then activate neutrophils and cause migration to inflammatory sites [19, 60]. IL-17C is also known to activate NF-κB and MAPK pathways after binding to IL-17RA/ IL-17RE receptors [36, 45]. IL-17C stimulates Th17 T-cells to increase synthesis of IL-17A/F and IL-22, and IL-17A strongly induces IL-17C in keratinocytes, forming an inflammatory loop [33]. IL-17E binds to IL-17RA/IL-17RB receptors and activates NF-κB, MAPK, and C/EBP, but differs from IL-17A in that it induces Th2-type immune responses and then suppresses Th-17 [53]. IL-17E promotes the migration of eosinophils, but not neutrophils [19].

3.2 Proliferation and differentiation

Mice overexpressing IL-17A in K14+ keratinocytes showed yellowish thickening of the epidermis, loss of the stratum granulosum, elongation of the dermal papillae, areas of hyperkeratosis and parakeratosis in the stratum corneum, and multiple neutrophilic abscesses, which are strikingly reminiscent of the epidermis of psoriasis [61]. In a three-dimensional epidermal model using normal human keratinocytes, IL-17A downregulated the expression of epidermal terminal differentiation markers such as filaggrin and loricrin after 5–6 days of incubation [22, 23] and caused the loss of the epidermal granular layer [22]. It has been reported that MAP17, whose expression is enhanced in normal human keratinocytes by IL-17A stimulation, downregulates mRNA expression of filaggrin [62]. On the other hand, there is a report that IL-17A did not downregulate mRNA of filaggrin and loricrin in the 3D epidermis 24 h after administration of IL-17A recombinant [63]. Similarly, in an experiment in which IL-17A recombinant was administered to confluent *in vitro* keratinocytes transfected with control and *TRAF3IP2* shRNA-expressing lentiviruses, mRNA expression was confirmed 24 h after recombinant administration, and at this point, there was still no significant difference in the terminal differentiation marker genes [64]. What these papers have in common is that they confirmed the expression of terminal differentiation marker genes after a short period of IL-17A stimulation (24 h), which may have led to different data from those reported above [22, 23]. In fact, another paper using *in vitro* normal human keratinocytes showed that mRNA expression of keratin 10, an intermediate differentiation marker gene, was downregulated at 48 h after IL-17A stimulation [65]. They also showed in the same paper that IL-17A promoted keratinocyte proliferation as well as IL-22 [65]. There are several reports that IL-17A promotes epidermal proliferation as shown in this paper. In particular, it is known that asymmetric cell divisions predominate over symmetric cell divisions in the epidermis of psoriasis, and IL-17A was reported to cause asymmetric cell divisions in normal human keratinocytes [66]. It has been reported that the promotion of epidermal proliferation by IL-17A is mediated by C/EBPα [67]. Furthermore, regenerating islet-derived protein 3-alpha (REG3A) promotes proliferation by suppressing keratinocyte terminal differentiation after injury to the epidermis [68]. In an experiment using a three-dimensional human epidermis model, IL-17C had no specific effect on the final differentiation markers of the epidermis and did not cause granular layer loss [22]. However, there is a report that overexpression of IL-17C in K5+ keratinocytes promotes psoriasiform skin inflammation in mouse experiments [34]. IL-17E promoted keratinocyte proliferation and upregulated the expression of keratin 10 in two- and three-dimensional cultures [69].

3.3 Migration

IL-17A increased actin stress fibers, promoted cellular contractility, and increased proteolytic collagen degradation, resulting in the increased migration potential of normal human keratinocytes [70]. A431, an epidermoid carcinoma cell line, also showed similar findings, suggesting that IL-17A may promote the invasion of malignant skin tumors. There are also reports of IL-17E affecting cell migration. This was accompanied by specific changes in the organization of the actin cytoskeleton and cell-substrate adhesion [69].

3.4 Adhesion

In a study using HaCaT cells, microarray experiments showed that IL-17A decreased the expression of 16 adhesion-related genes, including various integrins, plakoglobin, plakophilin, cadherin (including E-cadherin), claudin 7, and ZO-2 protein [71]. However, studies using normal human keratinocytes have not clarified the down regulation of adhesion molecules, and this needs to be verified in the future.

4. Skin diseases in which IL-17 is clearly involved

4.1 Psoriasis

The mRNA expression of IL-17A, IL-17C, IL-17E, and IL-17F is upregulated in psoriatic skin [51, 54, 72]. However, the fact that anti-IL-17A antibodies resolve the psoriatic skin lesions indicates that IL-17A is the most important player in the pathogenesis of psoriasis [1–3]. IL-17A induces inflammation and is directly involved in neutrophil migration and epidermal differentiation [19–22]. IL-17C stimulates Th17 T-cells to increase the synthesis of IL-17A/F and IL-22, and IL-17A strongly induces IL-17C in keratinocytes to form an inflammatory loop [33], but there are still insufficient human data indicating that IL-17C inhibition is effective in psoriasis. IL-17C may not be an essential component of IL-17A synthesis. In addition, there is no significant difference between the effects of IL-17RA antibodies (Brodalumab) and IL-17A antibodies (Ixekizumab and Secukinumab), which are the major receptors for IL-17A, IL-17E, and IL-17F. IL-17RA antibodies inhibit IL-17A and IL-17F signaling, but also block IL-17E signaling, so the possibility of offsetting Th17 suppression should be considered. If Bimekizumab, IL-17A/IL-17F antibodies, show greater efficacy than IL-17A in the future, the importance of IL-17F in psoriasis will be firmly established.

4.2 Hidradenitis suppurativa

The pathogenesis of Hidradenitis Suppurativa (HS) is still not fully understood [73]. HS is reported to be caused by follicular occlusion induced by keratosis and hyperplasia of the follicular epithelium, which eventually leads to the development of cysts [73, 74]. The occluded follicles eventually rupture, releasing their contents into the dermis, including keratin fibers, dermal detritus, multiple injuries, and pathogen-associated molecular patterns. Inflammatory immune pathways such as inflammasomes, Toll-like receptors, and IL-23/IL-17 signaling pathways are then activated [75]. The keratinocytes and innate immune cells activated will likely prompt a strong Th17 response, further activating the keratinocytes and recruiting neutrophils and other innate immune cells in an inflammatory loop. Th17 cells and neutrophils have been reported as the major IL-17-producing cells in the lesioned skin of HS [76, 77], and IL-17A gene expression in the lesioned skin of HS patients has been found to be 30- to 149-fold increased compared with healthy control skin [78, 79]. Recently, it has also been suggested that IL-17C may be involved in the pathogenesis of HS [80]. A variety of IL-17-related biologics are currently being investigated for HS, including CJM-112, secukinumab, bimekizumab, brodalumab, and guselkumab. Secukinumab treatment in HS has shown positive results in a series of case studies and open-label trials, with a Phase III trial underway [81, 82].

5. Conclusion

The effect of IL-17 on keratinocytes and the epidermis has a direct and significant impact not only on inflammation but also on epidermal differentiation. In particular, IL-17A has been the most studied, as clinical findings indicate that it is strongly involved in the pathogenesis of psoriasis and HS. However, other IL-17 family members have not been studied as extensively as IL-17A, and further basic research will provide new insights.

Conflict of interest

The authors declare no conflict of interest.

Author details

Emi Sato* and Shinichi Imafuku
Fukuoka University Faculty of Medicine, Department of Dermatology,
Fukuoka, Japan

*Address all correspondence to: emsato@fukuoka-u.ac.jp

IntechOpen

References

[1] Ratner M. IL-17-targeting biologics aim to become standard of care in psoriasis. Nature Biotechnology. 2015;**33**:3-4. DOI: 10.1038/nbt0115-3

[2] Imafuku S, Honma M, Okubo Y, Komine M, Ohtsuki M, Morita A, et al. Efficacy and safety of secukinumab in patients with generalized pustular psoriasis: A 52-week analysis from phase III open-label multicenter Japanese study. The Journal of Dermatology. 2016;**43**:1011-1017. DOI: 10.1111/1346-8138.13306

[3] Imafuku S, Torisu-Itakura H, Nishikawa A, Zhao F, Cameron GS, Japanese U-SG. Efficacy and safety of ixekizumab treatment in Japanese patients with moderate-to-severe plaque psoriasis: Subgroup analysis of a placebo-controlled, phase 3 study (UNCOVER-1). The Journal of Dermatology. 2017;**44**:1285-1290. DOI: 10.1111/1346-8138.13927

[4] Nikiphorou E, Buch MH, Hyrich KL. Biologics registers in RA: Methodological aspects, current role and future applications. Nature Reviews Rheumatology. 2017;**13**:503-510. DOI: 10.1038/nrrheum.2017.81

[5] Pouillon L, Travis S, Bossuyt P, Danese S, Peyrin-Biroulet L. Head-to-head trials in inflammatory bowel disease: Past, present and future. Nature Reviews Gastroenterology & Hepatology. 2020;**17**:365-376. DOI: 10.1038/s41575-020-0293-9

[6] Boguniewicz M. Biologic therapy for atopic dermatitis: Moving beyond the practice parameter and guidelines. The Journal of Allergy and Clinical Immunology in Practice. 2017;**5**:1477-1487. DOI: 10.1016/j.jaip.2017.08.031

[7] Lebwohl M. Psoriasis. Lancet. 2003;**361**:1197-1204. DOI: 10.1016/S0140-6736(03)12954-6

[8] Krueger JG. The immunologic basis for the treatment of psoriasis with new biologic agents. Journal of the American Academy of Dermatology. 2002;**46**:1-23. DOI: 10.1067/mjd.2002.120568

[9] Lowes MA, Bowcock AM, Krueger JG. Pathogenesis and therapy of psoriasis. Nature. 2007;**445**:866-873. DOI: 10.1038/nature05663

[10] Bernink JH, Ohne Y, Teunissen MBM, Wang J, Wu J, Krabbendam L, et al. c-Kit-positive ILC2s exhibit an ILC3-like signature that may contribute to IL-17-mediated pathologies. Nature Immunology. 2019;**20**:992-1003. DOI: 10.1038/s41590-019-0423-0

[11] Teunissen MBM, Munneke JM, Bernink JH, Spuls PI, Res PCM, Te Velde A, et al. Composition of innate lymphoid cell subsets in the human skin: Enrichment of NCR(+) ILC3 in lesional skin and blood of psoriasis patients. The Journal of Investigative Dermatology. 2014;**134**:2351-2360. DOI: 10.1038/jid.2014.146

[12] Cai Y, Shen X, Ding C, Qi C, Li K, Li X, et al. Pivotal role of dermal IL-17-producing gammadelta T cells in skin inflammation. Immunity. 2011;**35**:596-610. DOI: 10.1016/j.immuni.2011.08.001

[13] Paul C, Griffiths CEM, van de Kerkhof PCM, Puig L, Dutronc Y, Henneges C, et al. Ixekizumab provides superior efficacy compared with ustekinumab over 52 weeks of treatment: Results from IXORA-S, a phase 3 study. Journal of the American Academy of Dermatology. 2019;**80**(70-79):e73. DOI: 10.1016/j.jaad.2018.06.039

[14] Dyring-Andersen B, Honore TV, Madelung A, Bzorek M, Simonsen S, Clemmensen SN, et al. Interleukin (IL)-17A and IL-22-producing

neutrophils in psoriatic skin. The British Journal of Dermatology. 2017; **177**:e321-e322. DOI: 10.1111/bjd.15533

[15] Mashiko S, Bouguermouh S, Rubio M, Baba N, Bissonnette R, Sarfati M. Human mast cells are major IL-22 producers in patients with psoriasis and atopic dermatitis. The Journal of Allergy and Clinical Immunology. 2015;**136**(351-359):e351. DOI: 10.1016/j.jaci.2015.01.033

[16] Brembilla NC, Stalder R, Senra L, Boehncke WH. IL-17A localizes in the exocytic compartment of mast cells in psoriatic skin. The British Journal of Dermatology. 2017;**177**:1458-1460. DOI: 10.1111/bjd.15358

[17] Res PC, Piskin G, de Boer OJ, van der Loos CM, Teeling P, Bos JD, et al. Overrepresentation of IL-17A and IL-22 producing CD8 T cells in lesional skin suggests their involvement in the pathogenesis of psoriasis. PLoS One. 2010;**5**:e14108. DOI: 10.1371/journal.pone.0014108

[18] Ueyama A, Imura C, Fusamae Y, Tsujii K, Furue Y, Aoki M, et al. Potential role of IL-17-producing CD4/CD8 double negative alphabeta T cells in psoriatic skin inflammation in a TPA-induced STAT3C transgenic mouse model. Journal of Dermatological Science. 2017;**85**:27-35. DOI: 10.1016/j.jdermsci.2016.10.007

[19] Iwakura Y, Ishigame H, Saijo S, Nakae S. Functional specialization of interleukin-17 family members. Immunity. 2011;**34**:149-162. DOI: 10.1016/j.immuni.2011.02.012

[20] Cai XY, Gommoll CP Jr, Justice L, Narula SK, Fine JS. Regulation of granulocyte colony-stimulating factor gene expression by interleukin-17. Immunology Letters. 1998;**62**:51-58. DOI: 10.1016/s0165-2478(98)00027-3

[21] Hirai Y, Iyoda M, Shibata T, Kuno Y, Kawaguchi M, Hizawa N, et al. IL-17A stimulates granulocyte colony-stimulating factor production via ERK1/2 but not p38 or JNK in human renal proximal tubular epithelial cells. American Journal of Physiology. Renal Physiology. 2012;**302**:F244-F250. DOI: 10.1152/ajprenal.00113.2011

[22] Sato E, Yano N, Fujita Y, Imafuku S. Interleukin-17A suppresses granular layer formation in a 3-D human epidermis model through regulation of terminal differentiation genes. The Journal of Dermatology. 2020;**47**:390-396. DOI: 10.1111/1346-8138.15250

[23] Pfaff CM, Marquardt Y, Fietkau K, Baron JM, Luscher B. The psoriasis-associated IL-17A induces and cooperates with IL-36 cytokines to control keratinocyte differentiation and function. Scientific Reports. 2017; **7**:15631. DOI: 10.1038/s41598-017-15892-7

[24] Furue M, Furue K, Tsuji G, Nakahara T. Interleukin-17A and Keratinocytes in Psoriasis. International Journal of Molecular Sciences. 2020;**21**. DOI: 10.3390/ijms21041275

[25] Takatori H, Kanno Y, Watford WT, Tato CM, Weiss G, Ivanov II, et al. Lymphoid tissue inducer-like cells are an innate source of IL-17 and IL-22. The Journal of Experimental Medicine. 2009;**206**:35-41. DOI: 10.1084/jem.20072713

[26] Kolls JK, Linden A. Interleukin-17 family members and inflammation. Immunity. 2004;**21**:467-476. DOI: 10.1016/j.immuni.2004.08.018

[27] Li H, Chen J, Huang A, Stinson J, Heldens S, Foster J, et al. Cloning and characterization of IL-17B and IL-17C, two new members of the IL-17 cytokine family. Proceedings of the National Academy of Sciences of the United States of America. 2000;**97**:773-778. DOI: 10.1073/pnas.97.2.773

[28] Kouri VP, Olkkonen J, Ainola M, Li TF, Bjorkman L, Konttinen YT, et al. Neutrophils produce interleukin-17B in rheumatoid synovial tissue. Rheumatology (Oxford, England). 2014;**53**:39-47. DOI: 10.1093/rheumatology/ket309

[29] Reynolds JM, Lee YH, Shi Y, Wang X, Angkasekwinai P, Nallaparaju KC, et al. Interleukin-17B Antagonizes Interleukin-25-Mediated Mucosal Inflammation. Immunity. 2015;**42**:692-703. DOI: 10.1016/j.immuni.2015.03.008

[30] Ramirez-Carrozzi V, Ota N, Sambandam A, Wong K, Hackney J, Martinez-Martin N, et al. Cutting Edge: IL-17B Uses IL-17RA and IL-17RB to induce type 2 inflammation from human lymphocytes. Journal of Immunology. 2019;**202**:1935-1941. DOI: 10.4049/jimmunol.1800696

[31] Bie Q, Jin C, Zhang B, Dong H. IL-17B: A new area of study in the IL-17 family. Molecular Immunology. 2017;**90**:50-56. DOI: 10.1016/j.molimm.2017.07.004

[32] Bastid J, Dejou C, Docquier A, Bonnefoy N. The emerging role of the IL-17B/IL-17RB pathway in cancer. Frontiers in Immunology. 2020;**11**:718. DOI: 10.3389/fimmu.2020.00718

[33] Guttman-Yassky E, Krueger JG. IL-17C: A unique epithelial cytokine with potential for targeting across the spectrum of atopic dermatitis and psoriasis. The Journal of Investigative Dermatology. 2018;**138**:1467-1469. DOI: 10.1016/j.jid.2018.02.037

[34] Johnston A, Fritz Y, Dawes SM, Diaconu D, Al-Attar PM, Guzman AM, et al. Keratinocyte overexpression of IL-17C promotes psoriasiform skin inflammation. Journal of Immunology. 2013;**190**:2252-2262. DOI: 10.4049/jimmunol.1201505

[35] Vandeghinste N, Klattig J, Jagerschmidt C, Lavazais S, Marsais F, Haas JD, et al. Neutralization of IL-17C reduces skin inflammation in mouse models of psoriasis and atopic dermatitis. The Journal of Investigative Dermatology. 2018;**138**:1555-1563. DOI: 10.1016/j.jid.2018.01.036

[36] Nies JF, Panzer U. IL-17C/IL-17RE: Emergence of a unique axis in TH17 Biology. Frontiers in Immunology. 2020;**11**:341. DOI: 10.3389/fimmu.2020.00341

[37] Huang J, Meng S, Hong S, Lin X, Jin W, Dong C. IL-17C is required for lethal inflammation during systemic fungal infection. Cellular & Molecular Immunology. 2016;**13**:474-483. DOI: 10.1038/cmi.2015.56

[38] Conti HR, Whibley N, Coleman BM, Garg AV, Jaycox JR, Gaffen SL. Signaling through IL-17C/IL-17RE is dispensable for immunity to systemic, oral and cutaneous candidiasis. PLoS One. 2015;**10**:e0122807. DOI: 10.1371/journal.pone.0122807

[39] Song X, Zhu S, Shi P, Liu Y, Shi Y, Levin SD, et al. IL-17RE is the functional receptor for IL-17C and mediates mucosal immunity to infection with intestinal pathogens. Nature Immunology. 2011;**12**:1151-1158. DOI: 10.1038/ni.2155

[40] Pfeifer P, Voss M, Wonnenberg B, Hellberg J, Seiler F, Lepper PM, et al. IL-17C is a mediator of respiratory epithelial innate immune response. American Journal of Respiratory Cell and Molecular Biology. 2013;**48**:415-421. DOI: 10.1165/rcmb.2012-0232OC

[41] Wolf L, Sapich S, Honecker A, Jungnickel C, Seiler F, Bischoff M, et al. IL-17A-mediated expression of epithelial IL-17C promotes inflammation during acute pseudomonas aeruginosa pneumonia. American Journal of

Physiology. Lung Cellular and Molecular Physiology. 2016;**311**:L1015-L1022. DOI: 10.1152/ajplung.00158.2016

[42] Peng T, Chanthaphavong RS, Sun S, Trigilio JA, Phasouk K, Jin L, et al. Keratinocytes produce IL-17c to protect peripheral nervous systems during human HSV-2 reactivation. The Journal of Experimental Medicine. 2017; **214**:2315-2329. DOI: 10.1084/jem. 20160581

[43] Johansen C, Riis JL, Gedebjerg A, Kragballe K, Iversen L. Tumor necrosis factor alpha-mediated induction of interleukin 17C in human keratinocytes is controlled by nuclear factor kappaB. The Journal of Biological Chemistry. 2011;**286**:25487-25494. DOI: 10.1074/ jbc.M111.240671

[44] Chang SH, Reynolds JM, Pappu BP, Chen G, Martinez GJ, Dong C. Interleukin-17C promotes Th17 cell responses and autoimmune disease via interleukin-17 receptor E. Immunity. 2011;**35**:611-621. DOI: 10.1016/j. immuni.2011.09.010

[45] Ramirez-Carrozzi V, Sambandam A, Luis E, Lin Z, Jeet S, Lesch J, et al. IL-17C regulates the innate immune function of epithelial cells in an autocrine manner. Nature Immunology. 2011;**12**:1159-1166. DOI: 10.1038/ni.2156

[46] Liu X, Sun S, Liu D. IL-17D: A less studied cytokine of IL-17 family. International Archives of Allergy and Immunology. 2020;**181**:618-623. DOI: 10.1159/000508255

[47] Starnes T, Broxmeyer HE, Robertson MJ, Hromas R. Cutting edge: IL-17D, a novel member of the IL-17 family, stimulates cytokine production and inhibits hemopoiesis. Journal of Immunology. 2002;**169**:642-646. DOI: 10.4049/jimmunol.169.2.642

[48] Aggarwal S, Gurney AL. IL-17: Prototype member of an emerging cytokine family. Journal of Leukocyte Biology. 2002;**71**:1-8

[49] Broxmeyer HE, Sherry B, Cooper S, Lu L, Maze R, Beckmann MP, et al. Comparative analysis of the human macrophage inflammatory protein family of cytokines (chemokines) on proliferation of human myeloid progenitor cells. Interacting effects involving suppression, synergistic suppression, and blocking of suppression. Journal of Immunology. 1993;**150**:3448-3458

[50] Shin HC, Benbernou N, Esnault S, Guenounou M. Expression of IL-17 in human memory CD45RO+ T lymphocytes and its regulation by protein kinase A pathway. Cytokine. 1999;**11**:257-266. DOI: 10.1006/ cyto.1998.0433

[51] Johansen C, Usher PA, Kjellerup RB, Lundsgaard D, Iversen L, Kragballe K. Characterization of the interleukin-17 isoforms and receptors in lesional psoriatic skin. The British Journal of Dermatology. 2009;**160**:319-324. DOI: 10.1111/j.1365-2133.2008.08902.x

[52] Saenz SA, Siracusa MC, Perrigoue JG, Spencer SP, Urban JF Jr, Tocker JE, et al. IL25 elicits a multipotent progenitor cell population that promotes T(H)2 cytokine responses. Nature. 2010;**464**:1362-1366. DOI: 10.1038/nature08901

[53] Kleinschek MA, Owyang AM, Joyce-Shaikh B, Langrish CL, Chen Y, Gorman DM, et al. IL-25 regulates Th17 function in autoimmune inflammation. The Journal of Experimental Medicine. 2007;**204**:161-170. DOI: 10.1084/ jem.20061738

[54] Xu M, Lu H, Lee YH, Wu Y, Liu K, Shi Y, et al. An Interleukin-25-mediated autoregulatory circuit in keratinocytes plays a pivotal role in psoriatic skin inflammation. Immunity. 2018;**48** (787-798):e784. DOI: 10.1016/j.immuni. 2018.03.019

[55] Wright JF, Bennett F, Li B, Brooks J, Luxenberg DP, Whitters MJ, et al. The human IL-17F/IL-17A heterodimeric cytokine signals through the IL-17RA/IL-17RC receptor complex. Journal of Immunology. 2008;**181**:2799-2805. DOI: 10.4049/jimmunol.181.4.2799

[56] Gaffen SL. Structure and signalling in the IL-17 receptor family. Nature Reviews Immunology. 2009;**9**:556-567. DOI: 10.1038/nri2586

[57] Puel A, Cypowyj S, Bustamante J, Wright JF, Liu L, Lim HK, et al. Chronic mucocutaneous candidiasis in humans with inborn errors of interleukin-17 immunity. Science. 2011;**332**:65-68. DOI: 10.1126/science.1200439

[58] Dominguez-Villar M, Hafler DA. Immunology. An innate role for IL-17. Science. 2011;**332**:47-48. DOI: 10.1126/science.1205311

[59] Glatt S, Baeten D, Baker T, Griffiths M, Ionescu L, Lawson ADG, et al. Dual IL-17A and IL-17F neutralisation by bimekizumab in psoriatic arthritis: Evidence from preclinical experiments and a randomised placebo-controlled clinical trial that IL-17F contributes to human chronic tissue inflammation. Annals of the Rheumatic Diseases. 2018;**77**:523-532. DOI: 10.1136/annrheumdis-2017-212127

[60] Chiricozzi A, Nograles KE, Johnson-Huang LM, Fuentes-Duculan J, Cardinale I, Bonifacio KM, et al. IL-17 induces an expanded range of downstream genes in reconstituted human epidermis model. PLoS One. 2014;**9**:e90284. DOI: 10.1371/journal.pone.0090284

[61] Croxford AL, Karbach S, Kurschus FC, Wortge S, Nikolaev A, Yogev N, et al. IL-6 regulates neutrophil microabscess formation in IL-17A-driven psoriasiform lesions. The Journal of Investigative Dermatology.

2014;**134**:728-735. DOI: 10.1038/jid.2013.404

[62] Noh M, Yeo H, Ko J, Kim HK, Lee CH. MAP17 is associated with the T-helper cell cytokine-induced down-regulation of filaggrin transcription in human keratinocytes. Experimental Dermatology. 2010;**19**:355-362. DOI: 10.1111/j.1600-0625.2009.00902.x

[63] Rabeony H, Petit-Paris I, Garnier J, Barrault C, Pedretti N, Guilloteau K, et al. Inhibition of keratinocyte differentiation by the synergistic effect of IL-17A, IL-22, IL-1alpha, TNFalpha and oncostatin M. PLoS One. 2014;**9**:e101937. DOI: 10.1371/journal.pone.0101937

[64] Lambert S, Swindell WR, Tsoi LC, Stoll SW, Elder JT. Dual role of Act1 in keratinocyte differentiation and host defense: TRAF3IP2 silencing alters keratinocyte differentiation and inhibits IL-17 responses. The Journal of Investigative Dermatology. 2017;**137**:1501-1511. DOI: 10.1016/j.jid.2016.12.032

[65] Ekman AK, Bivik Eding C, Rundquist I, Enerback C. IL-17 and IL-22 promote keratinocyte stemness in the germinative compartment in psoriasis. The Journal of Investigative Dermatology. 2019;**139**(1564-1573):e1568. DOI: 10.1016/j.jid.2019.01.014

[66] Charruyer A, Fong S, Vitcov GG, Sklar S, Tabernik L, Taneja M, et al. Brief report: Interleukin-17A-dependent asymmetric stem cell divisions are increased in human psoriasis: A mechanism underlying benign hyperproliferation. Stem Cells. 2017;**35**:2001-2007. DOI: 10.1002/stem.2656

[67] Ma WY, Jia K, Zhang Y. IL-17 promotes keratinocyte proliferation via the downregulation of C/EBPalpha. Experimental and Therapeutic

Medicine. 2016;**11**:631-636.
DOI: 10.3892/etm.2015.2939

[68] Lai Y, Li D, Li C, Muehleisen B, Radek KA, Park HJ, et al. The antimicrobial protein REG3A regulates keratinocyte proliferation and differentiation after skin injury. Immunity. 2012;**37**:74-84. DOI: 10.1016/j.immuni.2012.04.010

[69] Borowczyk J, Buerger C, Tadjrischi N, Drukala J, Wolnicki M, Wnuk D, et al. IL-17E (IL-25) and IL-17A differentially affect the functions of human keratinocytes. The Journal of Investigative Dermatology. 2020;**140**:1379, e1372-1389. DOI: 10.1016/j.jid.2019.12.013

[70] Das S, Srinivasan S, Srivastava A, Kumar S, Das G, Das S, et al. Differential influence of IL-9 and IL-17 on actin cytoskeleton regulates the migration potential of human keratinocytes. Journal of Immunology. 2019;**202**:1949-1961. DOI: 10.4049/jimmunol.1800823

[71] Gutowska-Owsiak D, Schaupp AL, Salimi M, Selvakumar TA, McPherson T, Taylor S, et al. IL-17 downregulates filaggrin and affects keratinocyte expression of genes associated with cellular adhesion. Experimental Dermatology. 2012;**21**:104-110. DOI: 10.1111/j.1600-0625.2011.01412.x

[72] Senra L, Stalder R, Alvarez Martinez D, Chizzolini C, Boehncke WH, Brembilla NC. Keratinocyte-derived IL-17E contributes to inflammation in psoriasis. The Journal of Investigative Dermatology. 2016;**136**:1970-1980. DOI: 10.1016/j.jid.2016.06.009

[73] Fletcher JM, Moran B, Petrasca A, Smith CM. IL-17 in inflammatory skin diseases psoriasis and hidradenitis suppurativa. Clinical and Experimental Immunology. 2020;**201**:121-134. DOI: 10.1111/cei.13449

[74] von Laffert M, Helmbold P, Wohlrab J, Fiedler E, Stadie V, Marsch WC. Hidradenitis suppurativa (acne inversa): Early inflammatory events at terminal follicles and at interfollicular epidermis. Experimental Dermatology. 2010;**19**:533-537. DOI: 10.1111/j.1600-0625.2009.00915.x

[75] Vossen A, van der Zee HH, Prens EP. Hidradenitis suppurativa: A systematic review integrating inflammatory pathways into a cohesive pathogenic model. Frontiers in Immunology. 2018;**9**:2965. DOI: 10.3389/fimmu. 2018.02965

[76] Moran B, Sweeney CM, Hughes R, Malara A, Kirthi S, Tobin AM, et al. Hidradenitis suppurativa is characterized by dysregulation of the Th17: Treg cell axis, which is corrected by anti-TNF therapy. The Journal of Investigative Dermatology. 2017;**137**:2389-2395. DOI: 10.1016/j. jid.2017.05.033

[77] Lima AL, Karl I, Giner T, Poppe H, Schmidt M, Presser D, et al. Keratinocytes and neutrophils are important sources of proinflammatory molecules in hidradenitis suppurativa. The British Journal of Dermatology. 2016;**174**:514-521. DOI: 10.1111/bjd.14214

[78] Schlapbach C, Hanni T, Yawalkar N, Hunger RE. Expression of the IL-23/Th17 pathway in lesions of hidradenitis suppurativa. Journal of the American Academy of Dermatology. 2011;**65**:790-798. DOI: 10.1016/j.jaad.2010.07.010

[79] Kelly G, Hughes R, McGarry T, van den Born M, Adamzik K, Fitzgerald R, et al. Dysregulated cytokine expression in lesional and nonlesional skin in hidradenitis suppurativa. The British Journal of Dermatology. 2015;**173**:1431-1439. DOI: 10.1111/bjd.14075

[80] Navrazhina K, Frew JW, Krueger JG. Interleukin 17C is elevated in lesional

tissue of hidradenitis suppurativa. The British Journal of Dermatology. 2020;**182**:1045-1047. DOI: 10.1111/bjd.18556

[81] Casseres RG, Prussick L, Zancanaro P, Rothstein B, Joshipura D, Saraiya A, et al. Secukinumab in the treatment of moderate to severe hidradenitis suppurativa: Results of an open-label trial. Journal of the American Academy of Dermatology. 2020;**82**:1524-1526. DOI: 10.1016/j.jaad.2020.02.005

[82] Ribero S, Ramondetta A, Fabbrocini G, Bettoli V, Potenza C, Chiricozzi A, et al. Effectiveness of Secukinumab in the treatment of moderate-severe hidradenitis suppurativa: Results from an Italian multicentric retrospective study in a real-life setting. Journal of the European Academy of Dermatology and Venereology. 2021;**35**:e441-e442. DOI: 10.1111/jdv.17178

Chapter 4

Keratinocyte Stem Cells: Role in Aging

Tuba Musarrat Ansary, Koji Kamiya and Mamitaro Ohtsuki

Abstract

Stem cells located in the skin are responsible for continual regeneration, wound healing, and differentiation of different cell lineages of the skin. The three main locations of skin stem cells are the epidermis, dermis, and hair follicles. The keratinocyte stem cells are located in the epidermal basal layer (the interfollicular stem cells), hair follicle bulge region (the hair follicle stem cells), and sebaceous glands (the sebaceous gland stem cells) and are responsible for the epidermal proliferation, differentiation, and apoptosis. The interfollicular (IF) stem cells are responsible for epidermis regeneration by proliferating basal cells that attach to the underlying basement membrane and with time they exit from the cell cycle, start terminal differentiation, and move upward to form the spinous, the granular, and the stratum corneum layers. The hair follicle (HF) stem cells are responsible for hair regeneration and these stem cells undergo a cycle consists three stages; growth cycles (anagen), degeneration (catagen), and relative resting phase (telogen). The sebaceous gland (SG) stem cells located in between the hair follicle bulge and the gland and are responsible for producing the entire sebaceous gland which secretes oils to moisture our skin. The role of epidermal stem cells is extremely crucial because they produce enormous numbers of keratinocytes over a lifetime to maintain epidermal homeostasis. However, the age-associated changes in the skin; for example; alopecia, reduced hair density, gray or thin hair, reduced wound healing capacity are related to skin stem cells' decline functionality with age.

Keywords: interfollicular stem cells, hair follicle stem cells, melanocyte stem cells, chronological aging, photoaging

1. Introduction

Skin, the largest organ in the human body gives a protective barrier against the harmful events of the environment, for example, radiation, germs, temperature, and toxic substances. Moreover, the skin also protects our body from excessive dehydration and works as a permeability barrier. To support the above-mentioned roles and repair skin injury and wounds, the skin needs to regenerate and proliferate with the help of skin stem cells. Broadly the skin can be divided into three parts: Epidermis, the outermost layer, is mainly composed of keratinocytes and is known as the squamous stratified epithelium. The dermis, the middle layer, is separated by the basement membrane from the epidermis and is mainly composed of the extracellular matrix of tough collagen fibers, blood vessels, and nerves. The hypodermis, the third layer, mainly consists of fibroblasts, adipose tissues, and connective tissues. Adult skin development involves a multi-stage process that involved cells from diverse embryonic origins. Following gastrulation, the neuro-ectoderm cells stimulate the formation of the nervous system and skin epidermis. The neural and epidermal fate of these cells is dependent on different signaling pathways, for example, Wnt signaling, fibroblast growth factor (FGF), and bone morphogenetic protein (BMP) signaling, and Notch signaling pathway [1]. During development, the ectoderm-derived cells are become the epidermal basal layer and are responsible for all epidermal structures, for example, the hair follicles, sebaceous glands, and sweat glands [2]. A complex and multiple embryonic origins contribute to the dermis formation at different regions of the body. In a broad perspective, the mesoderm-derived cells are responsible for the dermis of the ventral and flank regions of the body and neural-crest-derived cells are responsible for the head dermis [2]. The mesoderm-derived cells are also responsible for the development of the adipose tissues in the hypodermis [3].

In everyday life skin has to perform many functions which are essential for our survival, for example, to protect from physical and mechanical injuries, harmful radiations, and after injury, it needs to form new cells to repair. In this regard, the skin stem cells (SCs) in the epidermis play the most essential roles to maintain skin renewal throughout life and repairing wounds after injury. In this review, we attempt to clarify the 1. Types of stem cells located in the epidermis, 2. The location, function, and markers to identify the epidermal SCs, 3. Role of chronological and photoaging on epidermal stem cells.

2. Epidermal stem cell

The epidermis is mainly consisting of five sub-layers with distinct characteristics although they function together to maintain tissue homeostasis and regeneration (**Figure 1**). The innermost or the deepest layer of the epidermis is the Basal layer, which is a single layer of proliferative basal cells that attach to a basement membrane (BM) by hemidesmosomes. These cells can continually divide and after some period lose attachment from the BM, get pushed by new cells and a program of differentiation has triggered. Above the basal layer, the Squamous cell layer contains the mature basal cells which are now known as squamous cells. This is the thickest layer of the epidermis and spiny projections that are attached to the surrounding cells by desmosomes. The keratinocytes are then got bigger and flatter and push towards two thin layers of the epidermis; the Stratum granulosum and the Stratum lucidum. The stratum corneum is the uppermost layer of the epidermis which contains the dead keratinocytes also referred to as anucleate squamous cells. The skin appendages include the sweat glands, sebaceous gland, nails, hair shafts, and hair follicles, and have both epidermal and dermal components which help the skin to complete its function [4]. The epidermal SCs reside in the basal layer are responsible for the maintenance of the epidermal stratified epithelium by self-renewal, wound repair, and differentiation capabilities. The epidermal proliferative unit (EPU) consists of basal cells that are responsible for the maintenance of the cornified layers by self-renewing and producing stem cells and non-stem cells [5]. There are several techniques available to understand and locate epidermal stem cells, for example, lineage tracing and genetic fate mapping. In the case of lineage tracing, an individual cell is labeled and the location, status, and the number of descendants from that cell can be identified by that label [6]. The genetic fate mapping technique involves marking an individual cell or a group of cells in the embryo and the trace the final position of all descendant cells until the completion of the development [7]. The epidermal SCs are also found in skin appendages, for example, the hair follicle bulge and the sebaceous glands, and have the potency to

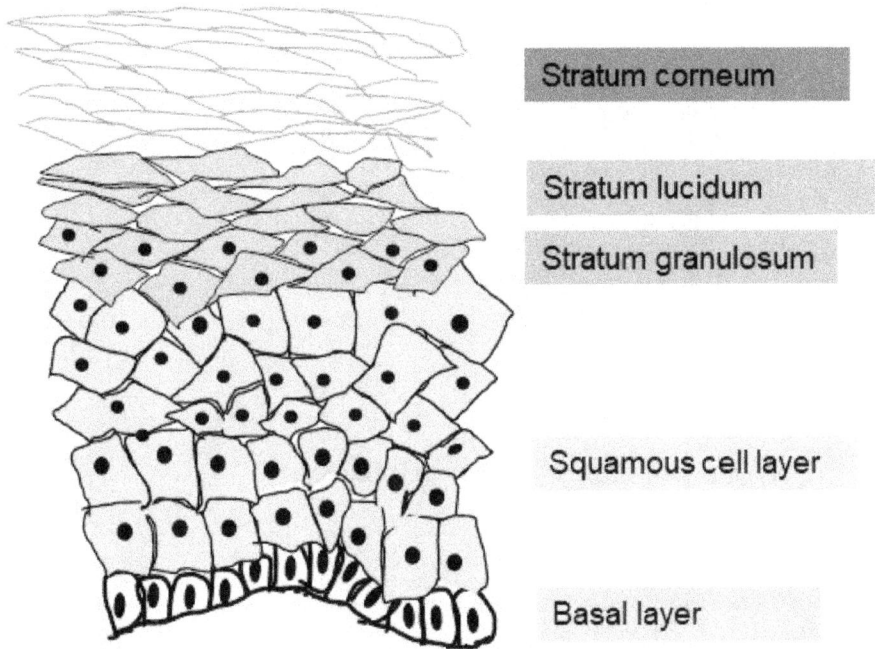

Figure 1.
The five main layers of epidermis. The basal cells can divide themselves and move upwards. As they move to the next layer they become flatter and start losing nuclei.

differentiate into different lineages that are present in adult skin [8]. The epidermal SCs niche can be classified into three groups according to their location; (1) The interfollicular (IF) stem cell located in the basal layer, (2) The hair follicle (HF) stem cell located in the bulge, and (3) The sebaceous gland (SG) stem cell located in between the bulge and hair shaft (**Figure 2**) [9, 10].

2.1 The interfollicular (IF) stem cell

The IF stem cells are found along with the basement membrane which is a specialized thin layer of extracellular matrices. The IF stem cells help to maintain the epidermis regeneration by self-renewal as well as produce progenitor cells named transit-amplifying (TA) cells which divide a limited number of times and then undergo terminal differentiation and give rise to flattened and dead keratinocytes in the cornified layer [11]. Apart from the multipotency these IF stem cells also show other characteristics, such as plasticity, which means these cells can lose their original identity and differentiate into other lineages [12]. Depending on the proximity from the wound and the specific stem cell niche origin, the epidermal stem cells participate in wound healing and tissue regeneration. By performing genetic fate mapping analysis, one report demonstrated that during initial wound healing there is an abundance of long-lived IF stem cells recruitment which promote re-epithelialization, and as the wound repairing the HF-derived stem cells increased in the wound area [13].

2.1.1 IF stem cell markers

There are several markers that can be used to identify IF stem cells (**Figure 3**). Integrins are the heterodimeric cell-surface receptors that consist of α and β subunits

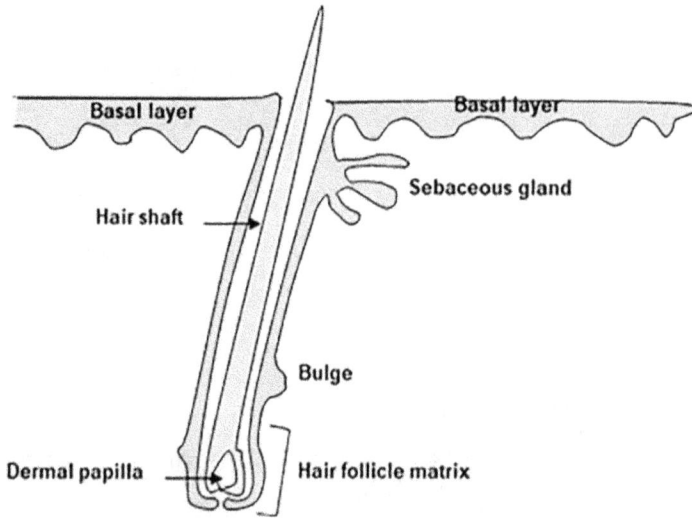

Figure 2.
Location of epidermal stem cells. The interfollicular (IF) stem cells located in the basal layer of the epidermis. Hair follicle (HF) stem cells located in the bulge area of the hair shaft. Melanocyte stem cells are found in the bulge and also in the hair matrix. The sebaceous gland (SG) stem cell located in the sebaceous gland.

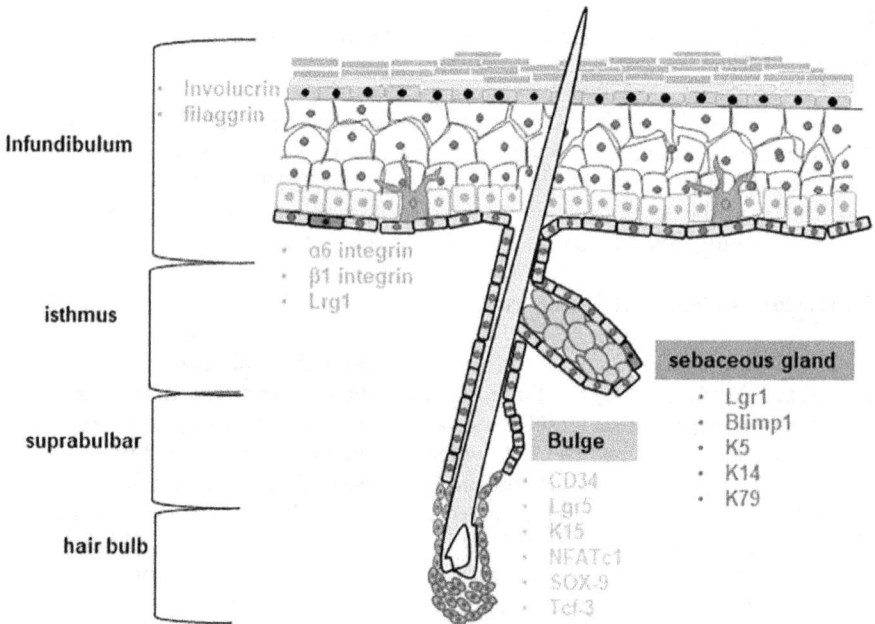

Figure 3.
Markers of epidermal stem cells.

and are responsible for cell adhesion, proliferation, and migration [14]. Adherence of the IF stem cells with the BM and extracellular matrixes is regulated by the integrins [15]. Several types of integrins are expressed in the epidermis; $\alpha 2\beta 1$ (receptor for Collagen), $\alpha 3\beta 1$ (receptor for laminin 5), $\alpha 6\beta 4$ (receptor for laminin), $\alpha 5\beta 1$ (receptor for fibronectin), [16, 17]. Among all integrins, the $\alpha 6$ or CD49f is the most used marker for epidermal stem cells [18]. The well-recognized marker for IF stem cells are the high α6 and the low transferrin receptor CD71 (α6-bright/CD71-dim) [19]. As there is a positive correlation between the IF stem cell proliferation and adherence, the proliferative IF stem cells can be distinguished from the low-adhered TA cells with the higher β1 integrins expression [20]. During terminal differentiation, the IF stem cells express involucrin, a differentiation marker, and filaggrin, an intermediate filament (IF)-associated protein [21, 22].

2.2 The hair follicle (HF) stem cell

The HF is one of the mini-organs in our body which go through life-long cyclic regeneration and involution [23–25]. The HF is located in attachment with the sebaceous gland and arrector pili muscle and it has two main segments; an epithelium made of keratinocytes and a dermal papilla (DP) made of mesenchymal cells [26, 27]. The cyclic regeneration of HF is mainly consisting of these phases; an active growth phase (anagen), a regression or involution phase (catagen), and a relative rest phase (telogen) and after the hair is shed a new hair cycle begins [28, 29]. The upper HF does not cycle visibly and is mainly divided into 2 segments; the infundibulum and the isthmus and the lower HF which consistently regenerates within the hair cycle divided into the hair bulb and the suprabulbar region [30, 31]. The infundibulum is the uppermost segment of the follicle which is funnel-shaped and begins from the epidermis surface and extends to the sebaceous gland opening [32]. The isthmus is the lower part of the upper HF and is located between the sebaceous gland and the bulge [27]. The bulb is the cyclic portion and the base of the HF which regenerates in every hair cycle and includes dermal papilla and HF matrix [28]. The suprabulbar region includes three parts; outer root sheath, inner root sheath, and the hair shaft and it lies between the hair bulb and the isthmus [33]. Bulge is the region where the HF stem cells are located and which lies between the sebaceous gland and the arrector pili muscle. These quiescent and long-lived stem cells have the potential to generate all epithelial lineages of the skin, including HF and hair [34–36]. The HF stem cells contribute to wound healing by recruiting multipotent SCs and life-long HF regeneration by providing new cells. They are normally known as quiescent, slow-cycling, and label-retaining cells.

Another type of stem cell that resides in the bulge of the HF is the long-lived neural-crest cell-derived melanocyte stem cell, which performs a crucial role in hair pigmentation maintenance [37]. Generally, melanoblasts, the immature progenitors of melanocytes, proliferate and differentiate into melanocytes in the epidermis and migrate to the hair follicle and divided into two categories; the hair matrix melanocytes responsible for pigmenting the original hair and the bulge melanocyte stem cells which are responsible for the following hair cycle pigmentation [38]. The regeneration of the follicular Melanocyte stem cells is synchronized with the HF cycle. During the anagen phase, the melanogenically active Melanocyte stem cells reside in the hair matrix proliferate and differentiate into Melanocyte progenitors to produce melanin and transfer to the neighboring keratinocytes and serve as a reservoir of the pigmentary unit for eye, hair, and skin [39]. In the catagen phase, the differentiated Melanocyte stem cells die because of the high apoptosis rate [40]. Melanin not only gives the essential pigmentation but also protects our skin from harmful UV radiation as the melanin granules work like a UV absorbent. To identify the lineage of the Melanocyte stem cell, a transgenic mouse model has been developed. The undifferentiated Melanocyte stem cells reside in the bulge express Dopachrome tautomerase (DCT) and tyrosinase-related proteins 1 (TRP-1) and serve as an early marker of Melanocyte stem cell. Nishimura et al. developed a transgenic mouse by using a lacz reporter manipulated by the DCT promoter and which enables people to find out the DCT positive melanoblasts [37]. However, both progenitor and mature melanocyte stem cells express DCT, so it cannot be regarded as a specific marker for the Melanocyte stem cell [41]. The CD34 can also use to identify the Melanocyte stem cells, for example, one paper reported that CD34 negative Melanocyte expressed high DCT, KIT (KIT Proto-Oncogene, Receptor Tyrosine Kinase), Tyr (tyrosinase), Tyrp1, Pmel17 (premelanosome protein), and MITF (Melanocyte Inducing Transcription Factor) [42]. Sox10 (Sry-related high-mobility-group box 10) can also be used as a marker for Melanocyte stem cells as this transcription factor plays an important role during the differentiation of the neural crest cell to Melanocyte stem cell [43, 44].

2.2.1 Location

There are some differences of opinion regarding the location of the stem cells and between the species. Some reports demonstrated that the germinative cells located in the lower area of the bulb are the HF stem cells as they have the differentiation ability [45, 46], however, several reports have challenged this idea and showed that HF stem cells are located in the bulge which is the upper and permanent portion of the HF. Several lines of experiments using pulse-chase experiment, in-vitro analysis, lineage analysis, seminal experiments, and BrdU- labeling experiments have proven the HF stem cells residing in the bulge [35, 46–51]. The in-vitro clonal analysis has shown that 95% of multipotent stem cells reside in the bulge and the rest of 5% are in the bulb, which is known as matrix cells or transit-amplifying (TA) cells [2, 52, 53].

2.2.2 Major functions

The HF stem cells located in the HF bulge area are label-retaining slow-cycling cells that perform several functions; for example, hair regeneration, reepithelization after a wound. HF stem cells play an important role in the generation of all layers of HF and hair regeneration by fueling the hair cycle [46, 48, 49, 54]. In general, when an anagen phase starts, the HF stem cells become activated and an HF will grow and regenerate and push the club hair above. During the anagen phase, the stem cells from the bulge area and the hair germ cells are activated by the mesenchymal cells from the DP, start proliferating in descending order, move to the bulb area, and create an outer root sheath (ORS) [55]. Throughout the anagen, the matrix or transit-amplifying (TA) cells originated from the bulge stem cells, move upward, and start to differentiate into follicle cells [49, 56]. By performing lineage tracing and double pulse-chase experiments one report confirmed that these TA cells then return to the bulge niche and lose the stemness property [55]. The catagen phase started when the matrix TA cells are exhausted and undergo apoptosis [47]. During catagen, apoptosis causes a huge decline in TA cell number, regression of approximately two-thirds of the hair follicles and only long-lived stem cells survive [57, 58]. After catagen, the resting phase or telogen phase will start. The telogen phase includes quiescent HF stem cells and shedding of the old HFs and this phase becomes longer progressively throughout life [57, 58]. In response to the signals from the DP, a new anagen phase started after the telogen phase and a new hair cycle begins [30, 59].

In addition to hair regeneration, the HF stem cells also play an important role in wound healing and re-epithelization. HF stem cells have the potential to differentiate into multiple lineages; for example, keratinocytes, smooth muscle cells, glial cells, neurons, and melanocytes and promote angiogenesis [60–63]. Many reports perform in-vitro and ex-vivo analysis using rodent and human samples showed that the HF stem cell can differentiate into audiogenic, osteogenic lineages as well as illustrate similar properties as bone-marrow-derived mesenchymal stem cells [64, 65]. Because of this property, the HF stem cells are regarded as one of the powerful stem cell candidates for cell therapy in the case of cutaneous wound healing and tissue regeneration [66]. In clinical studies, using graft transplantation from the scalp in patients with leg ulcers showed better therapeutic potential compared to the non-hairy grafts [67–70]. Performing double-label analysis and lineage tracing in the wound-repair model in animal reports showed that HF stem cells rapidly mobilize to the epidermis after injury and participate in epidermal repair by proliferating TA cells [36, 49, 71]. Using HF patch transplantation assay it has been demonstrated that HF stem cells contribute to generating new follicles in wounded mouse skin areas [72]. A complete reduction of HF stem cells in transgenic mice displayed a delay in wound healing after a full-thickness wound in the dorsal area [73]. Similarly, a delay in re-epithelization after the wound is observed in Edaraddcr/cr mice, that have a mutation in HF development [74]. Additionally, it has been shown that administration of HF stem cell in the wound area accelerates the healing process [75, 76].

2.2.3 HF stem cell markers

Several signaling pathways are important for the regulation and initiation of the anagen phase in the quiescent, slow-cycling label-retaining bulge stem cells [77]. Wnt/β catenin pathway plays an essential role in HF stem cells activation, maintenance, and differentiation [78]. The importance of Wnt/β catenin signaling in HF development is further proven by the report that showed complete HF follicle loss in a transgenic mouse with ectopic expression of Wnt inhibitor (Dickkopf 1) [79]. The fibroblast growth factor (FGF) signaling plays a crucial role in HF stem cell differentiation, hair cycle clock regulation, and angiogenesis [80, 81]. Sonic hedgehog (Shh) signaling expressed in the HF matrix is crucial for HF regeneration and neogenesis [82, 83]. Bone morphogenetic proteins (BMP) are also essential for HF regeneration, activation, quiescence, and TA cell differentiation and are expressed in the matrix [77, 84].

There are primarily four techniques to study skin stem cells; for example, label retention, clonogenic assays, skin reconstitution, and genetic lineage tracing [85]. Several markers have been identified to locate bulge and non-bulge stem cells in murine and human skin. Among the epithelial stem or progenitor markers, the most widely used marker for murine bulge stem cells are Keratin 15 (K15) and Clusters of differentiation 34 (CD34) [86–88]. In the case of human bulge stem cells, the most used markers are K15, Keratin 19 (K19), and Clusters of differentiation 200 (CD200) [89, 90]. The leucine-rich G protein–coupled receptor 5 (Lgr5), a Wnt target gene label the mouse lower bulge stem cells during the telogen phase and lower ORS HF during the anagen phase [91]. Several transcriptional factors are used to mark HF stem cells; such as Lim-homeodomain transcription factor Lhx2, SRY (Sex-determining region Y)-box 9 (Sox-9), transcription factor 3 (TCF-3), cytoplasmic 1 (NFATC1) (**Figure 3**) [92–94].

2.3 The sebaceous gland (SG) stem cell

Among other appendages in the skin, the sebaceous glands produce sebocytes and sebum to keep the lipid homeostasis and plays important role in barrier functions [95]. Unlike HF's cyclic growth, the SG has a continuous growth similar to the epidermis, and SG is typically found in association with the HF or as a modified version found independently in eyelids [96]. The resident stem cells in SG proliferate in the basal layer of the SG, differentiate into sebocytes and gather sebum, then move upwards and rupture the content inside into the pilosebaceous canal [97].

The specific markers that are used to identify the SG stem cells are K5, K14, K79, Leucine Rich Repeats, and Immunoglobulin Like Domains 1 (LRIG1), leucine-rich repeat-containing G protein-coupled receptor 1 (LGR6), B lymphocyte-induced maturation protein 1 (Blimp1) [98–101].

3. Epidermal stem cells and aging

Skin stem cells, residing in a protective niche, maintain the skin homeostasis by self-renewal and terminal differentiation. Unlike other somatic stem cells, the skin stem cells are quite resistant to aging as the number and self-renewal capacity of the stem cell do not reduce with age [102]. In general, stem cells stay at a quiescent state for a long time in their niche, and upon activation by numerous intrinsic and extrinsic factors these stem cells can exit this quiescent stage and differentiate into multiple lineages. Stem cell exhaustion is a state where the stem cells fail to renew themselves and thereby decrease in number which is mainly caused by aging. Several reports compared IF stem cells between young and old mice showed no difference in the number of stem cells, telomere length, gene expression related to aging, an abundance of K15 positive HF stem cells, and multipotency [103, 104]. On the contrary, by performing colony-forming essays in human keratinocytes, one report demonstrated that the cells from the aged donor have retarded colony-forming ability [105]. As we age there is an increase in senescent cells accumulation and DNA damage resulting in a decline in stem cells' function to produce new progenitor and effector cells [106, 107]. In this notion, Ultraviolet radiation (UVR) plays a crucial role in DNA damage in stem cells that ultimately lead to photoaging [108].

3.1 Photoaging and epidermal stem cell

The major characteristics of aged HF stem cells are imbalance in the phases of the hair cycle, stem cell exhaustion, and loss of hair (alopecia) and the appearance of the hair becoming dry, gray, or thin [109, 110]. A proper balance between the proliferation and quiescence state of the hair cycle is a prerequisite for HF stem cell lifespan and self-renewal. In this regard, the competitive balance between Wnt and BMP pathways is essential for HF homeostasis and cycle activation. During the regression phase, there is a decreased expression of Wnt and increased expression of BMP pathways which cause inhibition in keratinocyte proliferation and differentiation [111]. Specifically, the Wnt10 activated the anagen phase and BMP6 is the inhibitor of the hair cycle [77]. One report showed that persistent expression of Wnt1 causes mice HF to retain in the growth phase, initiate cellular senescence, and finally cause stem cell exhaustion and premature hair loss [112]. Increased Wnt signaling pathway, specially Wnt10 and β-catenin expression is observed in C57BL6/J mice exposed to UVR which causes HF miniaturization and gray hair [113]. UVR exposure can cause p53, a checkpoint protein, overactivation through DNA damage which is also associated with decreased stem cell renewal capacity, stem cell exhaustion, and premature aging [114]. The stem cell niche or microenvironment homeostasis is maintained by the interaction among mesenchymal cells, integrins expressed by the stem cells, and the extracellular matrix. It has been reported that UVR increased the expression of c-Myc, a transcriptional factor, which reduces the β1 integrin expression and thus impair β1 integrin-initiated adherence to the niche and migration [115–117]. Reactive oxidative species (ROS) induced by the UVR also cause a decrease in stem cell renewal capacity, senescence, and exhaustion [118, 119]. Another indication of stem cell aging is telomere shortening which resulted in hair loss and impaired stem cell proliferation and ultimately premature aging [120]. One report demonstrated chronic UV exposure to transgenic mice causes DNA damage and telomere shortening by modulating telomerase activity [121, 122]. A major hallmark of photoaged skin is altered wound repair capacity. Mitogen-activated protein kinase (MAPK) plays an essential role in cutaneous wound healing and an in-vivo study one report showed that chronic UV irradiation cause MAPK downregulation [123, 124].

3.2 Chronological aging and epidermal stem cell

A progressive decline in skin regeneration, repair, and homeostasis, thin hair, loss of hair, wrinkle, thin dermis, and epidermis, etc. are associated with accelerated aging. The major characteristic of the aged hair follicles is the hair density reduction and the resting period of the hair cycle increase. One report compared HF stem cell functions between 2 month and 24-month-old mice and found that the old mice HF showed a longer telogen phase, defective proliferation, and shorter hair growth phase [109]. Loss of HF stem cells is also associated with age-related hair shaft miniaturization [125]. Another typically aged phenotype related to hair is hair thinning, graying, or hair loss. There are several genetic mutation mice models that depicts accelerated aging phenotypes, for example, gray, brittle, and fragile hair and alopecia as a result of genetic instability [126]. It has been reported that DNA damage causes downregulation of Collagen 17 (Col17) expression in HF stem cells and these defected stem cells start to differentiate terminally and pushed themselves upward and eliminate [127]. As it is well known that Col17 is a crucial component in HF homeostasis and Col17 deficit can cause premature aging phenotype in hair, for example, alopecia or atrophy in HF [128]. In comparison to HF stem cells the role of aging on IF stem cells are not yet clarified. Some reports showed increased proliferation in epidermal stem cells and others confirmed the proliferation decreased as organisms aged. There are several proteins (Heat shock cognet 71, Stress protein 70, Myc associated protein, Cyclin D1, Glucose related protein) that expressed similarly in epidermal stem cells of adult and aged human skin and epidermal stem cells collected from the neonatal and aged mice have the same plasticity when injected into the blastocysts [129].

In normal physiological conditions, melanocytes, differentiated from the Melanocyte stem cell in the hair matrix, produce melanin during the anagen phase and transfer it to the neighboring keratinocytes. Following hair cycles, the TA melanocytes produced from the melanocyte stem cells are responsible for producing melanin and pigmenting new hair follicles. Several external or internal factors can cause aging-associated modifications in Melanocyte stem cells which cause gray hair, one of the most evident signs of aging. The genotoxic stress caused by radiation results in differentiation of the Mc stem cells in the niche and thereby diminish their self-renewal ability and which leads to hair pigmentation impairment in the following hair cycle [130]. Aging, itself is associated with Melanocyte stem cell reduction. An age-associated decline in Melanocyte stem cells measured by immunofluorescence with Kit antibody was observed in aged mice [109]. Another paper confirmed that the number of melanoblasts in the niche decreased with aging as well as the incidence of pigmented melanoblasts which means the ectopic differentiation increased with aging [131]. It can be speculated that due to the lack of enough progenitor cells the melanin production may be hampered and cause hair graying.

4. Conclusions

Skin stem cells participate in wound healing and maintain skin integrity and homeostasis by self-renewal and producing progenitor cells. Unlike other stem cells,

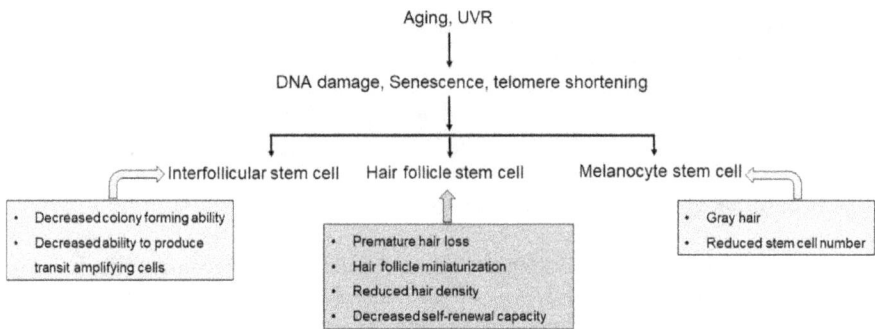

Figure 4.
Major effects of chronological and photo aging on different skin stem cells.

epidermal stem cells maintain an appropriate number throughout life and showed quite a resistance against aging. However, as we age the increased amount of DNA damage response and senescence can affect the epidermal stem cell's functions; for example, self-renewal capacity, increase exhaustion, mobility to the wound area or reduction in the number and that lead to skin aging phenotypes, for example, premature hair loss, gray or thin hair, reduced wound healing capacity (**Figure 4**).

Acknowledgements

I would like to express my sincere gratitude and acknowledgment to my supervisor and mentor Dr. Mayumi Komine for the guidance and supervision which help me to complete this project.

Conflict of interest

The author declares no conflict of interest.

Author details

Tuba Musarrat Ansary*, Koji Kamiya and Mamitaro Ohtsuki
Department of Dermatology, Jichi Medical University, Shimotsuke, Japan

*Address all correspondence to: tuba2020@jichi.ac.jp

IntechOpen

References

[1] Hu MS, Borrelli MR, Hong WX, Malhotra S, Cheung ATM, Ransom RC, et al. Embryonic skin development and repair. Organogenesis. 2018;**14**(1):46-63

[2] Blanpain C, Fuchs E. Epidermal stem cells of the skin. Annual Review of Cell and Developmental Biology. 2006; **22**:339-373

[3] Sebo ZL, Jeffery E, Holtrup B, Rodeheffer MS. A mesodermal fate map for adipose tissue. Development. 2018;**145**(17):1-11

[4] Yousef H, Miao JH, Alhajj M, Badri T. Histology, Skin Appendages. Treasure Island (FL): StatPearls; 2021

[5] Gonzalez-Celeiro M, Zhang B, Hsu YC. Fate by chance, not by choice: Epidermal stem cells go live. Cell Stem Cell. 2016;**19**(1):8-10

[6] Kretzschmar K, Watt FM. Lineage tracing. Cell. 2012;**148**(1-2):33-45

[7] Joyner AL, Zervas M. Genetic inducible fate mapping in mouse: Establishing genetic lineages and defining genetic neuroanatomy in the nervous system. Developmental Dynamics. 2006;**235**(9):2376-2385

[8] Weissman IL, Anderson DJ, Gage F. Stem and progenitor cells: Origins, phenotypes, lineage commitments, and transdifferentiations. Annual Review of Cell and Developmental Biology. 2001;**17**:387-403

[9] Fuchs E. Skin stem cells: Rising to the surface. The Journal of Cell Biology. 2008;**180**(2):273-284

[10] Qiu W, Chuong CM, Lei M. Regulation of melanocyte stem cells in the pigmentation of skin and its appendages: Biological patterning and therapeutic potentials. Experimental Dermatology. 2019;**28**(4):395-405

[11] Fuchs E, Raghavan S. Getting under the skin of epidermal morphogenesis. Nature Reviews. Genetics. 2002; **3**(3):199-209

[12] Kaur P. Interfollicular epidermal stem cells: Identification, challenges, potential. The Journal of Investigative Dermatology. 2006;**126**(7):1450-1458

[13] Vagnozzi AN, Reiter JF, Wong SY. Hair follicle and interfollicular epidermal stem cells make varying contributions to wound regeneration. Cell Cycle. 2015;**14**(21):3408-3417

[14] Rippa AL, Vorotelyak EA, Vasiliev AV, Terskikh VV. The role of integrins in the development and homeostasis of the epidermis and skin appendages. Acta Naturae. 2013; **5**(4):22-33

[15] Yang R, Wang J, Chen X, Shi Y, Xie J. Epidermal stem cells in wound healing and regeneration. Stem Cells International. 2020;**2020**:9148310

[16] Lee EC, Lotz MM, Steele GD Jr, Mercurio AM. The integrin alpha 6 beta 4 is a laminin receptor. The Journal of Cell Biology. 1992;**117**(3):671-678

[17] Watt FM. Role of integrins in regulating epidermal adhesion, growth and differentiation. The EMBO Journal. 2002;**21**(15):3919-3926

[18] Krebsbach PH, Villa-Diaz LG. The role of integrin alpha6 (CD49f) in stem cells: More than a conserved biomarker. Stem Cells and Development. 2017; **26**(15):1090-1099

[19] Terunuma A, Kapoor V, Yee C, Telford WG, Udey MC, Vogel JC. Stem cell activity of human side population and alpha6 integrin-bright keratinocytes defined by a quantitative in vivo assay. Stem Cells. 2007;**25**(3):664-669

[20] Jones PH, Watt FM. Separation of human epidermal stem cells from transit amplifying cells on the basis of differences in integrin function and expression. Cell. 1993;**73**(4):713-724

[21] Nair RP, Krishnan LK. Identification of p63+ keratinocyte progenitor cells in circulation and their matrix-directed differentiation to epithelial cells. Stem Cell Research & Therapy. 2013;**4**(2):38

[22] Forni MF, Ramos Maia Lobba A, Pereira Ferreira AH, Sogayar MC. Simultaneous isolation of three different stem cell populations from murine skin. PLoS One. 2015;**10**(10):e0140143

[23] Chase HB. Growth of the hair. Physiological Reviews. 1954;**34**(1): 113-126

[24] Chen CC, Plikus MV, Tang PC, Widelitz RB, Chuong CM. The modulatable stem cell niche: Tissue interactions during hair and feather follicle regeneration. Journal of Molecular Biology. 2016;**428**(7): 1423-1440

[25] Muller-Rover S, Handjiski B, van der Veen C, Eichmuller S, Foitzik K, McKay IA, et al. A comprehensive guide for the accurate classification of murine hair follicles in distinct hair cycle stages. The Journal of Investigative Dermatology. 2001;**117**(1):3-15

[26] Chen CL, Huang WY, Wang EHC, Tai KY, Lin SJ. Functional complexity of hair follicle stem cell niche and therapeutic targeting of niche dysfunction for hair regeneration. Journal of Biomedical Science. 2020;**27**(1):43

[27] Martel JL, Miao JH, Badri T. Anatomy, Hair Follicle. Treasure Island (FL): StatPearls; 2021

[28] Anzai A, Wang EHC, Lee EY, Aoki V, Christiano AM. Pathomechanisms of immune-mediated alopecia. International Immunology. 2019;**31**(7):439-447

[29] Xiao T, Yan Z, Xiao S, Xia Y. Proinflammatory cytokines regulate epidermal stem cells in wound epithelialization. Stem Cell Research & Therapy. 2020;**11**(1):232

[30] Schneider MR, Schmidt-Ullrich R, Paus R. The hair follicle as a dynamic miniorgan. Current Biology. 2009; **19**(3):R132-R142

[31] Harkey MR. Anatomy and physiology of hair. Forensic Science International. 1993;**63**(1-3):9-18

[32] Schneider MR, Paus R. Deciphering the functions of the hair follicle infundibulum in skin physiology and disease. Cell and Tissue Research. 2014;**358**(3):697-704

[33] Welle MM, Wiener DJ. The hair follicle: A comparative review of canine hair follicle anatomy and physiology. Toxicologic Pathology. 2016;**44**(4):564-574

[34] Inoue K, Aoi N, Sato T, Yamauchi Y, Suga H, Eto H, et al. Differential expression of stem-cell-associated markers in human hair follicle epithelial cells. Laboratory Investigation. 2009;**89**(8):844-856

[35] Morris RJ, Liu Y, Marles L, Yang Z, Trempus C, Li S, et al. Capturing and profiling adult hair follicle stem cells. Nature Biotechnology. 2004;**22**(4): 411-417

[36] Ito M, Liu Y, Yang Z, Nguyen J, Liang F, Morris RJ, et al. Stem cells in the hair follicle bulge contribute to wound repair but not to homeostasis of the epidermis. Nature Medicine. 2005;**11**(12):1351-1354

[37] Nishimura EK, Jordan SA, Oshima H, Yoshida H, Osawa M, Moriyama M, et al. Dominant role of the

niche in melanocyte stem-cell fate determination. Nature. 2002;**416**(6883):854-860

[38] Hirobe T, Enami H, Nakayama A. The human melanocyte and melanoblast populations per unit area of epidermis in the rete ridge are greater than in the inter-rete ridge. International Journal of Cosmetic Science. 2021;**43**(2):211-217

[39] Tobin DJ, Hagen E, Botchkarev VA, Paus R. Do hair bulb melanocytes undergo apoptosis during hair follicle regression (catagen)? The Journal of Investigative Dermatology. 1998;**111**(6):941-947

[40] Robinson KC, Fisher DE. Specification and loss of melanocyte stem cells. Seminars in Cell & Developmental Biology. 2009; **20**(1):111-116

[41] Lang D, Mascarenhas JB, Shea CR. Melanocytes, melanocyte stem cells, and melanoma stem cells. Clinics in Dermatology. 2013;**31**(2):166-178

[42] Joshi SS, Tandukar B, Pan L, Huang JM, Livak F, Smith BJ, et al. CD34 defines melanocyte stem cell subpopulations with distinct regenerative properties. PLoS Genetics. 2019;**15**(4):e1008034

[43] Harris ML, Buac K, Shakhova O, Hakami RM, Wegner M, Sommer L, et al. A dual role for SOX10 in the maintenance of the postnatal melanocyte lineage and the differentiation of melanocyte stem cell progenitors. PLoS Genetics. 2013;**9**(7):e1003644

[44] Marathe HG, Watkins-Chow DE, Weider M, Hoffmann A, Mehta G, Trivedi A, et al. BRG1 interacts with SOX10 to establish the melanocyte lineage and to promote differentiation. Nucleic Acids Research. 2017;**45**(11): 6442-6458

[45] Hardy MH. The secret life of the hair follicle. Trends in Genetics. 1992;**8**(2):55-61

[46] Oshima H, Rochat A, Kedzia C, Kobayashi K, Barrandon Y. Morphogenesis and renewal of hair follicles from adult multipotent stem cells. Cell. 2001;**104**(2):233-245

[47] Cotsarelis G, Sun TT, Lavker RM. Label-retaining cells reside in the bulge area of pilosebaceous unit: Implications for follicular stem cells, hair cycle, and skin carcinogenesis. Cell. 1990;**61**(7): 1329-1337

[48] Morris RJ, Potten CS. Highly persistent label-retaining cells in the hair follicles of mice and their fate following induction of anagen. The Journal of Investigative Dermatology. 1999;**112**(4):470-475

[49] Taylor G, Lehrer MS, Jensen PJ, Sun TT, Lavker RM. Involvement of follicular stem cells in forming not only the follicle but also the epidermis. Cell. 2000;**102**(4):451-461

[50] Panteleyev AA, Jahoda CA, Christiano AM. Hair follicle predetermination. Journal of Cell Science. 2001;**114**(Pt 19):3419-3431

[51] Rochat A, Kobayashi K, Barrandon Y. Location of stem cells of human hair follicles by clonal analysis. Cell. 1994;**76**(6):1063-1073

[52] Blanpain C, Lowry WE, Geoghegan A, Polak L, Fuchs E. Self-renewal, multipotency, and the existence of two cell populations within an epithelial stem cell niche. Cell. 2004;**118**(5):635-648

[53] Kobayashi K, Rochat A, Barrandon Y. Segregation of keratinocyte colony-forming cells in the bulge of the rat vibrissa. Proceedings of the National Academy of Sciences of the United States of America. 1993; **90**(15):7391-7395

[54] Tumbar T, Guasch G, Greco V, Blanpain C, Lowry WE, Rendl M, et al. Defining the epithelial stem cell niche in skin. Science. 2004;**303**(5656):359-363

[55] Hsu YC, Pasolli HA, Fuchs E. Dynamics between stem cells, niche, and progeny in the hair follicle. Cell. 2011;**144**(1):92-105

[56] Zhao J, Li H, Zhou R, Ma G, Dekker JD, Tucker HO, et al. Foxp1 regulates the proliferation of hair follicle stem cells in response to oxidative stress during hair cycling. PLoS One. 2015;**10**(7):e0131674

[57] Tamura Y, Takata K, Eguchi A, Kataoka Y. In vivo monitoring of hair cycle stages via bioluminescence imaging of hair follicle NG2 cells. Scientific Reports. 2018;**8**(1):393

[58] Houschyar KS, Borrelli MR, Tapking C, Popp D, Puladi B, Ooms M, et al. Molecular mechanisms of hair growth and regeneration: Current understanding and novel paradigms. Dermatology. 2020;**236**(4):271-280

[59] Lin KK, Kumar V, Geyfman M, Chudova D, Ihler AT, Smyth P, et al. Circadian clock genes contribute to the regulation of hair follicle cycling. PLoS Genetics. 2009;**5**(7):e1000573

[60] Amoh Y, Aki R, Hamada Y, Niiyama S, Eshima K, Kawahara K, et al. Nestin-positive hair follicle pluripotent stem cells can promote regeneration of impinged peripheral nerve injury. The Journal of Dermatology. 2012; **39**(1):33-38

[61] Joulai Veijouyeh S, Mashayekhi F, Yari A, Heidari F, Sajedi N, Moghani Ghoroghi F, et al. In vitro induction effect of 1,25(OH)2D3 on differentiation of hair follicle stem cell into keratinocyte. Biomedical Journal. 2017;**40**(1):31-38

[62] Babakhani A, Hashemi P, Mohajer Ansari J, Ramhormozi P, Nobakht M. In vitro differentiation of hair follicle stem cell into keratinocyte by simvastatin. Iranian Biomedical Journal. 2019;**23**(6):404-411

[63] Hoffman RM. The pluripotency of hair follicle stem cells. Cell Cycle. 2006;**5**(3):232-233

[64] Hoogduijn MJ, Gorjup E, Genever PG. Comparative characterization of hair follicle dermal stem cells and bone marrow mesenchymal stem cells. Stem Cells and Development. 2006;**15**(1):49-60

[65] Jahoda CA, Whitehouse J, Reynolds AJ, Hole N. Hair follicle dermal cells differentiate into adipogenic and osteogenic lineages. Experimental Dermatology. 2003;**12**(6):849-859

[66] Yari A, Heidari F, Veijouye SJ, Nobakht M. Hair follicle stem cells promote cutaneous wound healing through the SDF-1alpha/CXCR4 axis: An animal model. Journal of Wound Care. 2020;**29**(9):526-536

[67] Martinez ML, Escario E, Poblet E, Sanchez D, Buchon FF, Izeta A, et al. Hair follicle-containing punch grafts accelerate chronic ulcer healing: A randomized controlled trial. Journal of the American Academy of Dermatology. 2016;**75**(5):1007-1014

[68] Jimenez F, Garde C, Poblet E, Jimeno B, Ortiz J, Martinez ML, et al. A pilot clinical study of hair grafting in chronic leg ulcers. Wound Repair and Regeneration. 2012;**20**(6):806-814

[69] Tausche AK, Skaria M, Bohlen L, Liebold K, Hafner J, Friedlein H, et al. An autologous epidermal equivalent tissue-engineered from follicular outer root sheath keratinocytes is as effective as split-thickness skin autograft in recalcitrant vascular leg ulcers. Wound Repair and Regeneration. 2003;**11**(4): 248-252

[70] Ortega-Zilic N, Hunziker T, Lauchli S, Mayer DO, Huber C, Baumann Conzett K, et al. EpiDex(R) Swiss field trial 2004-2008. Dermatology. 2010;**221**(4):365-372

[71] Levy V, Lindon C, Zheng Y, Harfe BD, Morgan BA. Epidermal stem cells arise from the hair follicle after wounding. The FASEB Journal. 2007;**21**(7):1358-1366

[72] Biernaskie J, Paris M, Morozova O, Fagan BM, Marra M, Pevny L, et al. SKPs derive from hair follicle precursors and exhibit properties of adult dermal stem cells. Cell Stem Cell. 2009;**5**(6):610-623

[73] Chovatiya GL, Sarate RM, Sunkara RR, Gawas NP, Kala V, Waghmare SK. Secretory phospholipase A2-IIA overexpressing mice exhibit cyclic alopecia mediated through aberrant hair shaft differentiation and impaired wound healing response. Scientific Reports. 2017;**7**(1):11619

[74] Langton AK, Herrick SE, Headon DJ. An extended epidermal response heals cutaneous wounds in the absence of a hair follicle stem cell contribution. The Journal of Investigative Dermatology. 2008;**128**(5):1311-1318

[75] Babakhani A, Nobakht M, Pazoki Torodi H, Dahmardehei M, Hashemi P, Mohajer Ansari J, et al. Effects of hair follicle stem cells on partial-thickness burn wound healing and tensile strength. Iranian Biomedical Journal. 2020;**24**(2):99-109

[76] Heidari F, Yari A, Rasoolijazi H, Soleimani M, Dehpoor A, Sajedi N, et al. Bulge hair follicle stem cells accelerate cutaneous wound healing in rats. Wounds. 2016;**28**(4):132-141

[77] Wu P, Zhang Y, Xing Y, Xu W, Guo H, Deng F, et al. The balance of Bmp6 and Wnt10b regulates the telogen-anagen transition of hair follicles. Cell Communication and Signaling: CCS. 2019;**17**(1):16

[78] Alonso L, Fuchs E. Stem cells in the skin: Waste not, WNT not. Genes and Development. 2003;**17**(10): 1189-1200

[79] Andl T, Reddy ST, Gaddapara T, Millar SE. WNT signals are required for the initiation of hair follicle development. Developmental Cell. 2002;**2**(5):643-653

[80] Cai B, Zheng Y, Liu X, Yan J, Wang J, Yin G. A crucial role of fibroblast growth factor 2 in the differentiation of hair follicle stem cells toward endothelial cells in a STAT5-dependent manner. Differentiation. 2020;**111**:70-78

[81] Harshuk-Shabso S, Dressler H, Niehrs C, Aamar E, Enshell-Seijffers D. FGF and WNT signaling interaction in the mesenchymal niche regulates the murine hair cycle clock. Nature Communications. 2020;**11**(1):5114

[82] Lim CH, Sun Q, Ratti K, Lee SH, Zheng Y, Takeo M, et al. Hedgehog stimulates hair follicle neogenesis by creating inductive dermis during murine skin wound healing. Nature Communications. 2018;**9**(1):4903

[83] Woo WM, Zhen HH, Oro AE. Shh maintains dermal papilla identity and hair morphogenesis via a Noggin-Shh regulatory loop. Genes & Development. 2012;**26**(11):1235-1246

[84] Genander M, Cook PJ, Ramskold D, Keyes BE, Mertz AF, Sandberg R, et al. BMP signaling and its pSMAD1/5 target genes differentially regulate hair follicle stem cell lineages. Cell Stem Cell. 2014;**15**(5):619-633

[85] Kretzschmar K, Watt FM. Markers of epidermal stem cell subpopulations in adult mammalian skin. Cold Spring

Harbor Perspectives in Medicine. 2014;**4**(10):e013631

[86] Lyle S, Christofidou-Solomidou M, Liu Y, Elder DE, Albelda S, Cotsarelis G. The C8/144B monoclonal antibody recognizes cytokeratin 15 and defines the location of human hair follicle stem cells. Journal of Cell Science. 1998;**111**(Pt 21):3179-3188

[87] Dong G, Wang CL, Peng L, Ye L. Comparative study of cultivation of hair follicle bulge stem cell. Hua xi kou Qiang yi xue za zhi= Huaxi Kouqiang Yixue Zazhi= West China Journal of Stomatology. 2009;**27**(6):660-664

[88] Trempus CS, Morris RJ, Bortner CD, Cotsarelis G, Faircloth RS, Reece JM, et al. Enrichment for living murine keratinocytes from the hair follicle bulge with the cell surface marker CD34. The Journal of Investigative Dermatology. 2003;**120**(4):501-511

[89] Ohyama M, Terunuma A, Tock CL, Radonovich MF, Pise-Masison CA, Hopping SB, et al. Characterization and isolation of stem cell-enriched human hair follicle bulge cells. The Journal of Clinical Investigation. 2006;**116**(1): 249-260

[90] Pincelli C, Marconi A. Keratinocyte stem cells: Friends and foes. Journal of Cellular Physiology. 2010;**225**(2):310-315

[91] Jaks V, Barker N, Kasper M, van Es JH, Snippert HJ, Clevers H, et al. Lgr5 marks cycling, yet long-lived, hair follicle stem cells. Nature Genetics. 2008;**40**(11):1291-1299

[92] Nguyen H, Rendl M, Fuchs E. Tcf3 governs stem cell features and represses cell fate determination in skin. Cell. 2006;**127**(1):171-183

[93] Rhee H, Polak L, Fuchs E. Lhx2 maintains stem cell character in hair

follicles. Science. 2006;**312**(5782):1946-1949

[94] Horsley V, Aliprantis AO, Polak L, Glimcher LH, Fuchs E. NFATc1 balances quiescence and proliferation of skin stem cells. Cell. 2008;**132**(2):299-310

[95] Jang H, Myung H, Lee J, Myung JK, Jang WS, Lee SJ, et al. Impaired skin barrier due to sebaceous gland atrophy in the latent stage of radiation-induced skin injury: Application of non-invasive diagnostic methods. International Journal of Molecular Sciences. 2018;**19**(1):185

[96] Schneider MR, Paus R. Sebocytes, multifaceted epithelial cells: Lipid production and holocrine secretion. The International Journal of Biochemistry & Cell Biology. 2010;**42**(2):181-185

[97] Ghazizadeh S, Taichman LB. Multiple classes of stem cells in cutaneous epithelium: A lineage analysis of adult mouse skin. The EMBO Journal. 2001;**20**(6):1215-1222

[98] Saurat JH. Strategic targets in acne: The comedone switch in question. Dermatology. 2015;**231**(2):105-111

[99] Veniaminova NA, Grachtchouk M, Doane OJ, Peterson JK, Quigley DA, Lull MV, et al. Niche-specific factors dynamically regulate sebaceous gland stem cells in the skin. Developmental Cell. 2019;**51**(3):326-40e4

[100] Fullgrabe A, Joost S, Are A, Jacob T, Sivan U, Haegebarth A, et al. Dynamics of Lgr6(+) progenitor cells in the hair follicle, sebaceous gland, and interfollicular epidermis. Stem Cell Reports. 2015;**5**(5):843-855

[101] Horsley V, O'Carroll D, Tooze R, Ohinata Y, Saitou M, Obukhanych T, et al. Blimp1 defines a progenitor population that governs cellular input to the sebaceous gland. Cell. 2006;**126**(3):597-609

[102] Racila D, Bickenbach JR. Are epidermal stem cells unique with respect to aging? Aging (Albany NY). 2009;**1**(8):746-750

[103] Stern MM, Bickenbach JR. Epidermal stem cells are resistant to cellular aging. Aging Cell. 2007;**6**(4):439-452

[104] Giangreco A, Qin M, Pintar JE, Watt FM. Epidermal stem cells are retained in vivo throughout skin aging. Aging Cell. 2008;**7**(2):250-259

[105] Barrandon Y, Green H. Three clonal types of keratinocyte with different capacities for multiplication. Proceedings of the National Academy of Sciences of the United States of America. 1987;**84**(8):2302-2306

[106] Sharpless NE, DePinho RA. How stem cells age and why this makes us grow old. Nature Reviews Molecular Cell Biology. 2007;**8**(9):703-713

[107] Sperka T, Wang J, Rudolph KL. DNA damage checkpoints in stem cells, ageing and cancer. Nature Reviews Molecular Cell Biology. 2012;**13**(9): 579-590

[108] Panich U, Sittithumcharee G, Rathviboon N, Jirawatnotai S. Ultraviolet radiation-induced skin aging: The role of DNA damage and oxidative stress in epidermal stem cell damage mediated skin aging. Stem Cells International. 2016;**2016**:7370642

[109] Keyes BE, Segal JP, Heller E, Lien WH, Chang CY, Guo X, et al. Nfatc1 orchestrates aging in hair follicle stem cells. Proceedings of the National Academy of Sciences of the United States of America. 2013;**110**(51): E4950-E4959

[110] Goodier M, Hordinsky M. Normal and aging hair biology and structure 'aging and hair'. Current Problems in Dermatology. 2015;**47**:1-9

[111] Plikus MV, Mayer JA, de la Cruz D, Baker RE, Maini PK, Maxson R, et al. Cyclic dermal BMP signalling regulates stem cell activation during hair regeneration. Nature. 2008;**451**(7176): 340-344

[112] Castilho RM, Squarize CH, Chodosh LA, Williams BO, Gutkind JS. mTOR mediates Wnt-induced epidermal stem cell exhaustion and aging. Cell Stem Cell. 2009;**5**(3):279-289

[113] Zhai X, Gong M, Peng Y, Yang D. Effects of UV induced-photoaging on the hair follicle cycle of C57BL6/J Mice. Clinical, Cosmetic and Investigational Dermatology. 2021;**14**:527-539

[114] Lakin ND, Jackson SP. Regulation of p53 in response to DNA damage. Oncogene. 1999;**18**(53):7644-7655

[115] Takahashi S, Pearse AD, Marks R. The acute effects of ultraviolet-B radiation on c-myc and c-Ha ras expression in normal human epidermis. Journal of Dermatological Science. 1993;**6**(2):165-171

[116] Alarcon-Vargas D, Tansey WP, Ronai Z. Regulation of c-myc stability by selective stress conditions and by MEKK1 requires aa 127-189 of c-myc. Oncogene. 2002;**21**(28):4384-4391

[117] Waikel RL, Kawachi Y, Waikel PA, Wang XJ, Roop DR. Deregulated expression of c-Myc depletes epidermal stem cells. Nature Genetics. 2001; **28**(2):165-168

[118] Kwon SH, Park KC. Antioxidants as an epidermal stem cell activator. Antioxidants (Basel). 2020;**9**(10):958

[119] Zhou D, Shao L, Spitz DR. Reactive oxygen species in normal and tumor stem cells. Advances in Cancer Research. 2014;**122**:1-67

[120] Flores I, Cayuela ML, Blasco MA. Effects of telomerase and telomere

length on epidermal stem cell behavior. Science. 2005;**309**(5738):1253-1256

[121] Stout GJ, Blasco MA. Telomere length and telomerase activity impact the UV sensitivity syndrome xeroderma pigmentosum C. Cancer Research. 2013;**73**(6):1844-1854

[122] Ventura A, Pellegrini C, Cardelli L, Rocco T, Ciciarelli V, Peris K, et al. Telomeres and telomerase in cutaneous squamous cell carcinoma. International Journal of Molecular Sciences. 2019;**20**(6):1333

[123] Gong M, Zhang P, Li C, Ma X, Yang D. Protective mechanism of adipose-derived stem cells in remodelling of the skin stem cell niche during photoaging. Cellular Physiology and Biochemistry. 2018;**51**(5):2456-2471

[124] Thuraisingam T, Xu YZ, Eadie K, Heravi M, Guiot MC, Greemberg R, et al. MAPKAPK-2 signaling is critical for cutaneous wound healing. The Journal of Investigative Dermatology. 2010;**130**(1):278-286

[125] Xie Y, Chen D, Jiang K, Song L, Qian N, Du Y, et al. Hair shaft miniaturization causes stem cell depletion through mechanosensory signals mediated by a Piezo1-calcium-TNF-alpha axis. Cell Stem Cell. 2022;6;**29**(1):70-85.e6

[126] Hasty P, Campisi J, Hoeijmakers J, van Steeg H, Vijg J. Aging and genome maintenance: Lessons from the mouse? Science. 2003;**299**(5611):1355-1359

[127] Matsumura H, Mohri Y, Binh NT, Morinaga H, Fukuda M, Ito M, et al. Hair follicle aging is driven by transepidermal elimination of stem cells via COL17A1 proteolysis. Science. 2016;**351**(6273):aad4395

[128] Tanimura S, Tadokoro Y, Inomata K, Binh NT, Nishie W, Yamazaki S, et al. Hair follicle stem cells provide a functional niche for melanocyte stem cells. Cell Stem Cell. 2011;**8**(2):177-187

[129] Liang L, Chinnathambi S, Stern M, Tomanek-Chalkley A, Manuel TD, Bickenbach JR. As epidermal stem cells age they do not substantially change their characteristics. The Journal of Investigative Dermatology. Symposium Proceedings. 2004;**9**(3):229-237

[130] Sikkink SK, Mine S, Freis O, Danoux L, Tobin DJ. Stress-sensing in the human greying hair follicle: Ataxia Telangiectasia Mutated (ATM) depletion in hair bulb melanocytes in canities-prone scalp. Scientific Reports. 2020;**10**(1):18711

[131] Nishimura EK, Granter SR, Fisher DE. Mechanisms of hair graying: Incomplete melanocyte stem cell maintenance in the niche. Science. 2005;**307**(5710):720-724

Chapter 5

Keratinocytes in Skin Disorders: The Importance of Keratinocytes as a Barrier

Mayumi Komine, Jin Meijuan, Miho Kimura-Sashikawa,
Razib MD. Hossain, Tuba M. Ansary, Tomoyuki Oshio,
Jitlada Meephansan, Hidetoshi Tsuda, Shin-ichi Tominaga
and Mamitaro Ohtsuki

Abstract

Keratinocytes are the major structural component of the epidermis. They differentiate from the basal through spinous to granular layers, and abrupt loss of nucleus pushes them to differentiate into cornified layers, which exfoliates as scales. Differentiation process is tightly controlled by the organized expression of transcription factors and other regulators, which sustains the physiological function of the skin barrier. The genetic abnormality of the molecules expressed in this pathway causes hereditary skin disorders and defects in barrier function. Ichthyosis is caused by keratins, enzymes, and structural proteins involved in lipid metabolism and cornified envelope formation. Atopic dermatitis seemed to be an immune-oriented disease, but the recent finding revealed filaggrin as a causative factor. Keratinocytes respond to acute injury by releasing alarmins. IL-33 is one of such alarmins, which provoke Th2-type inflammation. IL-33 works as a cytokine and, at the same time, as nuclear protein. IL-33 has double-faced nature, with pro- and anti-inflammatory functions. Epidermis, covering the entire body, should stay silent at minor insults, while it should provoke inflammatory signals at emergency. IL-33 and other double-faced molecules may play a role in fine tuning the complexed function of epidermal keratinocytes to maintain the homeostasis of human body.

Keywords: keratinocytes, keratin, mutation, ichthyosis, hereditary skin disorders

1. Introduction

Keratinocytes are the principal epidermal cells constituting the outermost layer of the skin—the external and largest organ of the human body. They are immunologically active in that they produce various cytokines and chemokines, stimulating dendritic cells and lymphocytes to trigger inflammatory skin diseases, as well as they respond to cytokines produced from immune cells to establish skin lesions of inflammatory skin diseases, such as psoriasis and atopic dermatitis. They are also very efficient in avoiding harsh environmental assaults, such as chemical, mechanical, radiological, and microbial insults. The keratinocytes protect the dermal homeostasis by having a constant turnover whereby the basal (inner) layer differentiates into the cornified (outer) layer. Thus, they form a constant and perfect outer barrier to the inner dermal layers and the body. They also form a rigid mechanical barrier by cornification—constructing a brick-and-mortar type of structure with cornified cells and lipids, the defects in either of which cause hereditary skin disorders upon mutation. They also secrete various antimicrobial peptides, such as cathelicidin, psoriasin, defensin, and many S100 proteins, to protect the skin from infection. The nuclei of keratinocytes contain various alarmins, such as HMGB1, IL-33, and IL-1alpha, which can induce rapid and strong inflammation upon injury, but also can get promptly inactivated by the enzymes present in the inflammatory environment. However, malfunctioning of the keratinocytes at its immunological level or at a genetic/protein level can lead to pathological conditions such as psoriasis, atopic dermatitis, and hereditary skin disorders.

Keratinocytes, as the main component of outermost epidermal layer, should provoke and at the same time stop inflammation at appropriate time points to maintain a stable and healthy condition of not only the skin, but also the entire body. Keratinocytes harbor anti-inflammatory properties more than other types of cells do, such as lymphocytes, macrophages, and dendritic cells, as keratinocytes are always exposed to environmental insults. The mechanism of developing inflammatory conditions has been intensely investigated; however, the mechanism of ceasing inflammation has not been fully investigated. I speculate that a novel approach to maintaining healthy conditions would be unraveled when the mechanism of sequestrating inflammation is investigated and that epidermal keratinocytes are good candidates to investigate these mechanisms because they present pro- and anti-inflammatory properties in vivo and in vitro.

In this chapter, various cutaneous disorders have been discussed with emphasis on keratinocyte function and roles in pathogenesis. We have surveyed PubMed with each disease name, picked up the original literature with pivotal findings, reviewed articles covering the related area of interest, and wrote this chapter.

2. Epidermal keratinocytes

Epidermal keratinocytes form a stratified epithelium, consisting of basal, spinous, granular, and cornified layers starting from the dermal side. Epidermal keratinocytes exert their functions through structural components such as actin, microtubules, keratin filaments, desmosomes, hemidesmosomes, tight junctions, and adherence junctions; their motility, proliferation, and cytokine production being controlled by these structural proteins. Epidermal keratinocytes gradually differentiate through the layers—from the basal, spinous, and granular, ultimately to the cornified cell layer. They demonstrate various characteristic features owing to their differential function and according to their differentiation state, which are sometimes more complex than those of simple epithelial cells constituting the digestive tract and glands [1].

The primary and most important function of epidermal keratinocytes is their role as a physical barrier of the skin, in addition to their role as a responder to the external stimuli. The cornified cells, together with inter-cornified cell lipids, form cornified cell barriers to protect the inner body from harsh external environmental stimuli. The cornified cells, upon catalysis by transglutaminase 1, form a cornified cell envelope—a strong structure composed of filaggrin that aggregates keratin filaments, with various protein components such as involucrin, loricrin, SPR, and desmosomal proteins. Defects in the enzymes and protein components essential in forming cornified cell envelopes cause skin barrier dysfunction, resulting in skin disorders [2, 3].

Recent findings have revealed that some patients with atopic dermatitis (AD) harbor loss-of-function mutations in the filaggrin gene, resulting in severe skin barrier defects. Ichthyosis vulgaris (IV) is also caused by mutations in the filaggrin gene, but patients with this mutation develop either AD, IV, or both, indicating that mutations in the filaggrin gene alone are not enough to determine the phenotypes [4–7]. Mutations in the transglutaminase 1 gene and other genes important in the cornification processes, such as ATP-binding cassette subfamily A member 12 (ABCA12), and arachidonate 12-lipoxygenase 12 s type (ALOX12), cause hereditary ichthyosis, also known as acquired recessive congenital ichthyosis [8]. Connexin is a component of the gap junction, and mutation in gap junction protein beta 3 (GJB3) gene, which encodes connexin (Cx31) causes erythrokeratodermia variabilis, in which inflammatory erythematous eruptions with hyperkeratinization gradually changes its form [9]. Mutations in the loricrin gene cause loricrin keratoderma, with characteristic finger constriction ring formation or congenital ichthyosiform erythroderma [10, 11].

Steroid sulfatase is an enzyme that catalyzes the degradation of cholesterol sulfate, a molecule that functions in the attachment of cornified cells. The mutation in its gene causes X-lined ichthyosis, with retarded detachment of cornified cells, termed as retention hyperkeratosis. Point mutations result in typical skin manifestations; whereas, mutations spanning bigger lengths of this gene involving the surrounding genomic region are accompanied by other syndromic symptoms, such as mental retardation, short stature, and epilepsy [12].

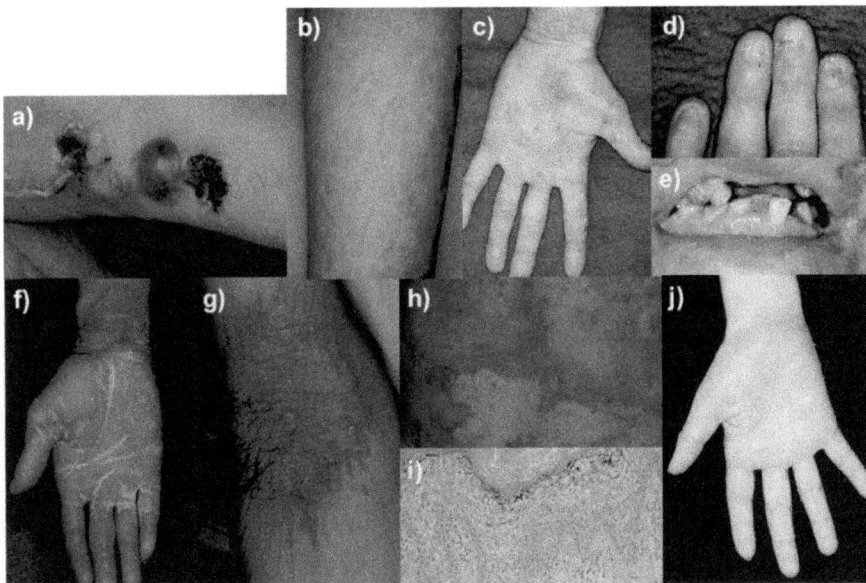

Figure 1.
Characteristic skin manifestation in hereditary skin disorders with mutation in genes expressed in keratinocytes. a) Bulla formation on the foot of a child having epidermolysis bullosa simplex (EBS) and with mutation in the KRT5 gene. b) Macular brownish pigmentation in EBS with mottled pigmentation in patients with mutation in KRT5 gene. c) Hyperkeratosis in hands, d) nail deformity and e) dental decay in patients with dystrophic EB and with mutation in the integrin beta 4 gene. f) Diffuse hyperkeratosis in hands, with lichenification in wrist, g) hyperkeratosis with lichenification in cubital fossa, h) small bulla formation on diffuse erythema and i) its histopathology with hematoxylin and eosin staining in patients with Epidermolytic hyperkeratosis and mutation in KRT1 gene.

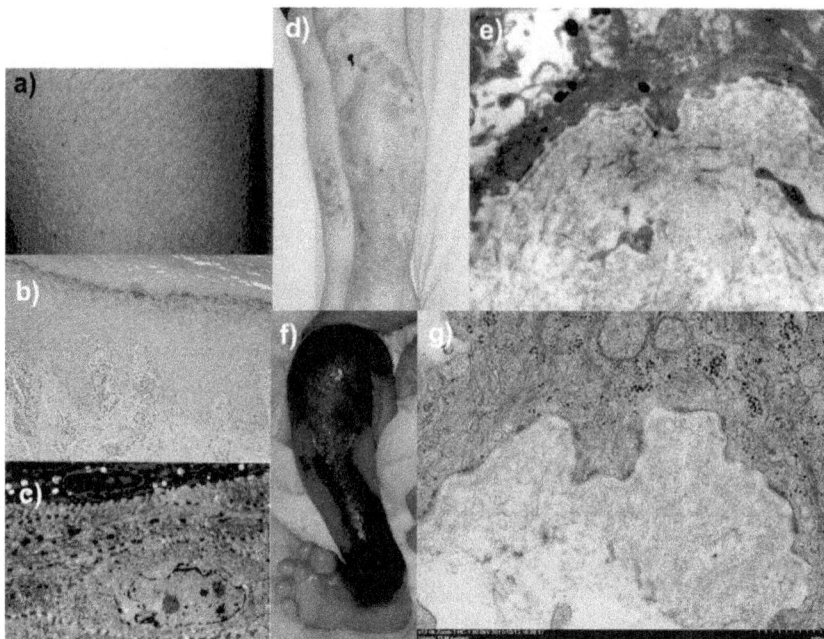

Figure 2.
Characteristic skin manifestation and electron microscopic findings of genetic skin disorders. a) Clinical manifestation of X-liked ichthyosis (XI) patient, b) histopathology with hematoxylin and eosin staining, c) electron microscopic features in patients with XI and deletion in STS gene. d) Clinical manifestation of autosomal dominant type dystrophic EB (AD-DEB) patient, e) electron microscopy of skin sample from AD-DEB patient with heterozygous mutation in collagen type VII (COL7). f) Clinical manifestation of autosomal recessive type dystrophic EB patient, g) electron microscopy of skin sample AR-DEB patient with homozygous mutation in COL7.

I.	**Common Ichthyosis**		
	Ichthyosis vulgaris	AD	FLG
	X-linked ichthyosis	XR	STS
II.	**Autosomal recessive congenital ichthyosis**		
	Harlequin ichthyosis	AR	ABCA12
	Lamellar ichthyosis	AR	TGM1; NIPAL4; ALOX12B; ABCA12
	Congenital ichthyosiform erythroderma	AR	ALOXE3; ALOX12B; ABCA12; NIPAL4; TGM1 CYP4F22
III.	**Keratinopathic ichthyosis**		
	Epidermolytic ichthyosis	AD	KRT1; KRT10
	Superficial epidermolytic ichthyosis	AD	KRT2
IV.	**Others**		
	Loricrin keratoderma	AD	LOR
V.	**X-linked ichthyosis syndrome**		
	X-linked ichthyosis syndromic presentation	XR	STS and others
	FPAP syndrome	XR	MBTPS2
VI.	**Autosomal ichthyosis syndrome**		
	Netherton syndrome	AR	SPINK5

Sjogren Larsson syndrome	AR	ALDH3A2
Refsum syndrome	AR	PHYH; PEX7
Gaucher syndrome II	AR	GBA
KID syndrome	AD	GJB2; GJB6

Table 1.
Representative non-syndromic and syndromic ichthyosis with causative genes. Modified citation from Ref. [3].

Some hereditary keratinizing disorders are accompanied by syndromic symptoms other than skin manifestations. Mutations in GJB2 gene, encoding Cx26, cause KID syndrome—with keratitis, ichthyosis, and deafness as triads, exhibiting papillomatous and spinous keratotic eruptions on the face and extremities, and with palmoplantar keratoderma and alopecia [13]. Mutation in the serine protease inhibitor SPINK5 causes Netherton syndrome, with atopic dermatitis-like skin eruptions, characteristic ichthyosis linearis circumflexa and bamboo hair [14]. Sjogren-Larsson syndrome is caused by a mutation in the fatty aldehyde dehydrogenase (ALDH3A2) gene, with clinical symptoms including ichthyosis, spastic limb paralysis, and mental retardation [15]. **Figures 1** and **2** shows skin manifestations of several hereditary skin disorders. **Table 1** shows a summary of the types of ichthyosis and their accompanying gene mutations.

3. Keratinopathies

Keratins are the main intermediate filaments of epidermal keratinocytes. The keratin family consists of more than 50 members; acidic keratin and basic keratin monomers pair to form heterodimers, which are then organized into tetramers with an anti-parallel alignment. Tetramers of keratins are stored at the peripheral boundaries of cells for filament formation when needed. Lateral and longitudinal aggregations of these tetramers and octamers form keratin filaments. Local pH, osmotic conditions, and phosphorylation status are thought to be the driving forces of filament formation [16].

Simple epithelia consist of simple epithelial keratins, such as K8, K18, and K19. Basal cells of the simple and stratified epithelia express K5 and K14, while the suprabasal cells express K1 and K10 in the interfollicular epidermis, K3 and K12 in the corneal epithelium, K4 and K13 in the esophageal epithelium, and K6 and K16 in the oral epithelium. The follicular epidermis and palmoplantar epidermis express K6, K16, and K17. Their expression is tightly controlled by transcription factors in a differentiation- and localization-dependent manner.

Mutations in keratin genes cause various hereditary skin disorders [17]. Mutations either in *KRT5* and *KRT14* gene cause epidermolysis bullosa simplex (EBS), manifested by bulla formation with slight mechanical forces from childhood. *KRT1* or *KRT10* gene mutations cause epidermolytic ichthyosis (EI), with characteristic histopathological features called epidermolytic hyperkeratosis characterized by large droplets of keratohyalin granules with vacuolization and hyperkeratosis in epidermal keratinocytes. A similar mutation in the *KRT9* gene causes Voerner-type palmoplantar keratoderma (PPK), with similar epidermolytic hyperkeratosis on the palm and soles, owing to the exclusive *KRT9* expression and distribution on palms and soles in humans. Similarly, a mutation in the *KRT2e*, expression of which is distributed in the granular layer of the epidermis, causes superficial epidermolytic ichthyosis. Pachyonychia congenita, manifested by thickening of finger and toenails and sometimes accompanying steatocystoma multiplex, is caused by mutations in *KRT6*, *KRT16*, or *KRT17*, which is expressed in nails and follicular epithelium [18]. White sponge nevus usually seen in the oral epithelium is caused by mutations in the *KRT4* or *KRT13* gene, with whitish, somewhat keratinized oral epithelium showing papillomatous growth [19]. Simple epithelial keratins, such as *KRT7*, *KRT8*, *KRT18*, and *KRT19*, are distributed not only in cutaneous glandular structures, such as sweat glands and sebaceous glands but also in various internal organs, including the digestive tract and liver. Mutations of these simple epithelial keratins in skin disorders have not yet been elucidated, but the importance of *KRT8* and *KRT18* mutations in liver diseases have been postulated. End-stage liver disease patients have been reported to show higher rates of *KRT 8/18* mutations [20]. The solubility of keratins depends on their phosphorylation status, and mutations in the phosphorylation site affect the solubility of keratin filaments, resulting in cell damage. Recent findings revealed that the phosphorylation of keratins is also affected by the acetylation or methylation status of keratins; thus, mutations at these sites also cause cell damage. Mutations in *KRT8* and *KRT18* affect the keratin phosphorylation, acetylation, or methylation, in turn, affecting the stability in keratin filaments, resulting in an imbalance between *KRT8* and *KRT18* proteins, and causing excessive oxidative stress and susceptibility to liver disorders [21, 22].

4. Adherence machinery of epidermal keratinocytes

Adherence machinery is indispensable for controlling keratinocyte cell motility, proliferation, and viability, as well as the epidermal barrier function by controlling cell attachment and cell tension. Keratinocytes have six major adherence mechanisms [23, 24]: 1) Hemidesmosomes, which connect basal keratinocytes to the dermal component, with cytoskeletal molecules such as keratins, 2) desmosomes, which connect neighboring keratinocytes, sustain the epidermal sheet structure and maintain tension by connecting to cytoskeletal molecules, such as keratins, 3) Adherence junctions, which control keratinocyte motility by connecting intracellular actin to E-cadherin in the adherence junctions to neighboring keratinocytes [25], 4) Gap junctions, which also have ion-transporter functions, indirectly control keratinocyte barrier function, 5) Tight junctions, which control the liquid interface in epithelia and consist of claudins and occludins [26], and 6) Focal adhesion—attachment of plaques connecting cells to the extracellular matrix, thereby, making connections to scaffolds to maintain the keratinocyte motility, proliferation, and viability.

Hemidesmosomal proteins are indispensable for maintaining normal dermalepidermal structures (**Figure 3**) [27, 28]. Mutations in hemidesmosomal protein genes, such as integrin alpha 6 or beta 4, cause junctional epidermolysis bullosa [29]. Mutations in plectin, a constituent of desmosomes and hemidesmosomes, cause junctional type epidermolysis bullosa with pyloric atresia [30]. Collagen type VII localizes from just beneath lamina densa to support attachment of lamina densa to the dermal structure. Mutations in the collagen type VII gene cause dystrophic epidermolysis bullosa with prominent skin ulcer and scar formation [31, 32]. These severe EBs usually occur in patients with homozygous mutation or compound heterozygous mutations. A heterozygous mutation, the same mutation but which harbors on only one allele of the gene, causes a milder form of EB, leading to the development of nodular prurigo-like lesions or scar formations in autosomal dominant type dystrophic EB or development of palmoplantar keratoderma with alopecia and dental deformation in autosomal dominant type junctional EB. A

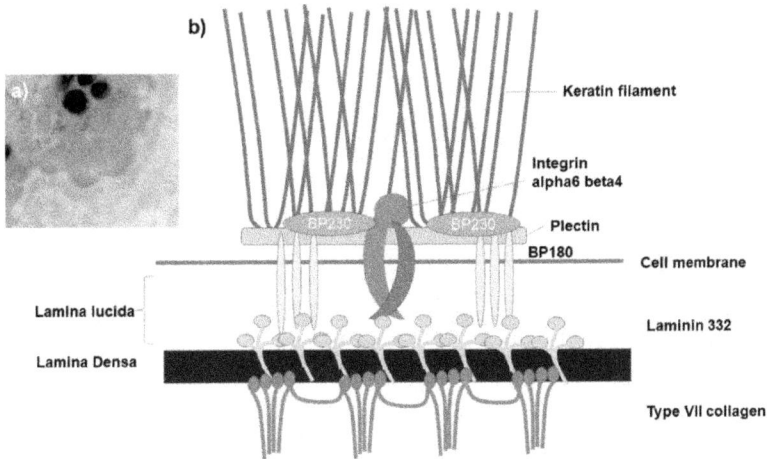

Figure 3.
The structure of hemidesmosome. a) Electron microscopy of hemidesmosomes and basal lamina of a normal human subject. b) Schematic view of a hemidesmosome structure. Plectin forms a platform where keratin filaments and hemidesmosomal proteins bind, crosslinking keratin filaments with integrin beta4. Transmembrane protein bullous pemphigoid antigen 1 (BP180) connects hemidesmosomes to laminin 332, a component of the lamina densa. Bullous pemphigoid antigen 2 (BP230) is an intra-cytoplasmic protein that composes the hemidesmosome from inside the cells.

recent study revealed that mutations in desmoplakin cause lethal acantholytic epidermolysis bullosa [33].

5. Skin barrier dysfunction in diseases

Atopic dermatitis patients present decreased filaggrin and ceramide contents in their cornified layers, along with decreased skin barrier function [34]. This dysfunctional barrier allows allergens to penetrate the skin, thus, resulting in sensitization to environmental allergens [35]. Peanut allergies are often observed in infants of families that consume large amounts of peanut and have detectable levels of peanut debris in the surrounding environment [36]. Exercise-induced food allergy also develops in relation to filaggrin mutation [37]. A previous study has shown that infants with frequent emollient hydration of skin showed a lower rate of bronchial asthma development compared to babies without emollient hydration of skin, indicating the importance of the skin barrier functioning in maintaining overall health and stable homeostasis [38]. **Figure 4** illustrates the epidermal structure with barrier proteins.

Mutations in filaggrin and protease inhibitors can cause atopic dermatitis. Netherton syndrome is caused by a mutation in the serine protease inhibitor KAZAL type5 (*SPINK5*), which results in atopic dermatitis-like skin eruption [39]. Nagashima-type PPK is caused by a mutation in *SERPINB7*, a serine protease inhibitor, with low-grade hyperkeratosis on the palms and soles, involving the backside of the fingers, toes, and a triangular lesion on the wrist. Some cases of Nagashima-type PPK also develop food allergies or atopic dermatitis [40]. Protease inhibitors are essential for stopping the catalyzing reaction by proteases, thus, protecting the skin barrier from over-degradation. The precise mechanisms underlying the development of atopic dermatitis or allergy in Nagashima-type PPK patients are not clear, but one theory is that proteinase activation receptors are potent pro-inflammatory

a)   b)

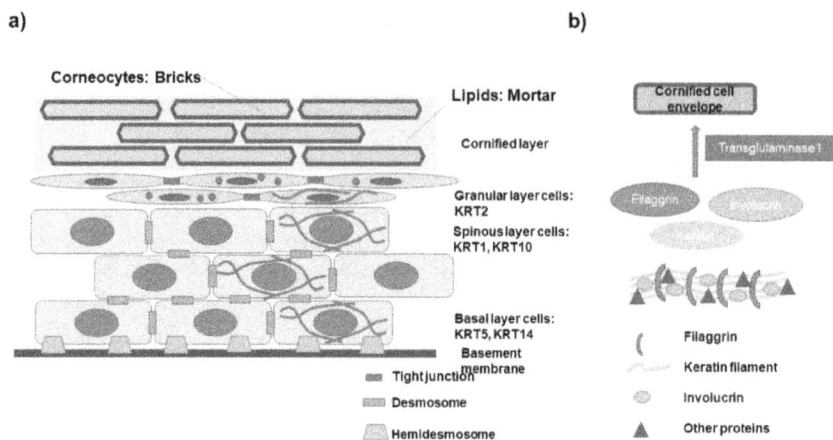

Figure 4.
The structure of epidermis and its adhesion molecules. a) Schematic view of epidermis structure and its adhesion molecules. Basal keratinocytes attach to basement membrane through hemidesmosomes, having keratins such as KRT5 and KRT14. Suprabasal keratinocytes start to produce KRT1 and KRT10 and attach to neighboring keratinocytes with desmosomes. Granular layer cells express KRT2 containing keratohyalin granules and attach to the neighboring cells with tight junctions, which also have desmosomes. Corneocytes lose nuclei and embed in lipid layers, connecting each other with corneodesmosomes. Cornified cell envelope develops when cells become corneal layer cells from granular layer cells. Adherence junctions exist from basal keratinocytes to granular layer keratinocytes. b) Cornified envelope development from its components. Filaggrin aggregates keratin filaments and involucrin, and other cornified envelope proteins gather to form a cornified cell envelope, upon catalysis by transglutaminase 1.

molecules that react with proteinases to induce inflammation. Thus, protease inhibitors could be a therapeutic target for atopic dermatitis [41].

Lipid abnormalities could be another cause of atopic dermatitis, demonstrating the skin barrier dysfunction. A decrease in ceramide content of the cornified layer has been demonstrated in patients with atopic dermatitis, which is another cause of skin barrier dysfunction [42]. Ceramide constitutes almost 50% of the lipids in the corneal layer and is indispensable for skin barrier function. Mutations in genes involved in lipid metabolism are not known in atopic dermatitis, but gene metabolic diseases, such as Gauche disease and Nieman-Pick disease that are characterized by mutations in the glucocerebrosidase and sphingomyelin phosphodiesterase 1 gene, respectively, which are indispensable in ceramide synthesis, resulting in development of atopic dermatitis-like skin eruptions from early childhood. Abnormalities in lipid metabolism could be another cause of atopic dermatitis, which requires further investigation [43].

Skin barrier function is affected not only by genetic conditions but also by ordinary routines of daily life. People who often scrub too much during bathing, bathe for a long time period or very frequently, and especially those who scrub their skin with nylon towels or scrubbing brushes show very dry skin with small scales all over the body. These individuals often complain of severe itching, especially after bathing, often resulting in eczema development. Excessive use of detergent also causes barrier disruption by increasing the pH of the skin, resulting in enhanced enzymatic activity of proteinases in the cornified layers [44]. These lifestyle routines would exacerbate eczematous changes in individuals having a genetic predisposition that makes them more susceptible to barrier disruption.

Keratinocytes form not only mechanical barriers but also chemical or immuno-logical barriers for humans. They express several antimicrobial peptides, such as cathelicidin, defensin, psoriasin, and various S100 proteins. These antimicrobial peptides prevent pathogenic microbes from colonizing the skin surface, thus conferring resistance to microbial infections [45]. Certain conditions such as atopic dermatitis have decreased production of antimicrobial peptides, leaving the individuals more susceptible to bacterial or viral infections through the skin [46, 47]. Filaggrin mutations are at times the direct cause of barrier disruption, but T helper (Th)2-skewed immune conditions can be another cause too, as Th2 type cytokines cause a less differentiated state of keratinocytes thus resulting in lower production of antimicrobial molecules and barrier proteins [48]. Mutations in filaggrin also cause ichthyosis vulgaris, which often co-exists in atopic dermatitis patients. However, not all patients with atopic dermatitis have ichthyosis vulgaris and vice versa, even in the presence of filaggrin gene mutations [7]. Thus, the filaggrin mutation alone cannot explain the pathogenesis of atopic dermatitis.

Psoriasis is another major inflammatory skin disorder that shows hyperproduction of antimicrobial peptides in the epidermis induced by skewed Th17 populations thus making patients resistant to skin infections [49, 50]. Cathelicidin—one of the antimicrobial peptides, complexes with self RNA or DNA to induce the activation of myeloid dendritic cells and plasmacytoid dendritic cells, respectively. This activation further triggers psoriatic inflammation, thus creating a positive feedback loop in pathogenesis of psoriasis [51, 52]. Recent advancements in translational research produced many "biologics", which target inflammatory cytokines, such as IL-17, TNF, and IL-23, as a treatment option for psoriasis. These include anti-TNF antibodies (adalimumab [53], infliximab [54], certolizumab-pegol [55]), anti-IL-17 antibodies (secukinumab [56, 57], ixekizumab [58], brodalumab [59], bimekizumab [60]), anti-IL-12/23p40 antibody (ustekinumab [61, 62]), and anti-IL-23p19 antibodies (guselkumab [63], risankizumab [64, 65], thildrakizumab [66–69]). Janus kinases (JAKs) are important intracellular signaling molecules downstream of cytokine receptors [70]. They are also deeply involved in inflammation in psoriasis, and JAK inhibitors have been developed as therapeutic options in psoriasis [71–76].

Ichthyosis has also been shown to have a Th17-skewed immune balance [77], and Th17 is a potent inducer of antifungal immunity. However, ichthyosis patients often develop cutaneous superficial fungal infections [78, 79]. Taken together, this suggests that the immune imbalance by itself cannot explain the susceptibility to fungal infections, meanwhile also implicating the importance of proper functioning of the skin barrier to avoid superficial fungal infections.

6. Danger signals from keratinocytes

As a barrier, keratinocytes respond to emergency conditions by releasing danger-associated molecular patterns (DAMPs) when acutely injured. IL-33 is one such emergency molecule, which resides in the nucleus in a steady-state, but is released when cells undergo necrosis to stimulate immune reactions [80]. IL-33 is a relatively recently identified member of the IL-1 family and functions mainly as a pro-inflammatory molecule, although under certain conditions, it can also work as an anti-inflammatory molecule. IL-1 alpha—the prototype of IL-1 family members, was identified as an alarmin several decades ago. The Koebner phenomenon in psoriasis is attributed to the release of IL-1 alpha from damaged keratinocytes, which induces psoriasis in regions after skin injury [81]. IL-1 alpha is an interesting cytokine that mainly functions as a soluble cytokine, but also shows a nuclear presence. It has been reported that IL-1 alpha repeatedly travels between the cytoplasm and nucleus, and is released into the extracellular space upon cell damage to provoke inflammation [82]. IL-33 has similar characteristics in that it resides in the nucleus too, is released during cell necrosis, and induces inflammation. IL-33, similar to IL-1 beta, is produced as a full-length pro-form. IL-33 pro-form is active, but even more, activated when digested by neutrophil elastases or cathepsin. It is, however, inactivated when digested by caspases, unlike IL-1 beta, which is activated by caspases during activation of NRLP3 inflammasomes. ST2L—a receptor of IL-33, is expressed on Th2 cells, group 2 innate lymphoid cells, and regulatory T cells, and its soluble form—sST2, blocks the interaction of IL-33 with ST2L [83].

IL-33 exhibits both pro-inflammatory and anti-inflammatory roles. As a Th2 cytokine, it stimulates ST2L-expressing cells, including mast cells, Th2, and ILC2 cells. This enhances Th2 type inflammation by inducing expression of Th2 type cytokines, such as IL-5 and IL-13. However, upon Th1 or Th17 activation, IL-33 may attenuate pathological conditions by skewing Th2 type inflammation. The graft versus host disease (GVHD) reaction [84] was reported to be attenuated by IL-33, and experimental autoimmune encephalomyelitis showed reduction in response to IL-33 action [85]. Graft rejection in heart transplantation was reported to be attenuated by treatment with IL-33 [86]. IL-33, by inducing regulatory T cell function, was shown to induce immunosuppression [87]. UVB-induced immunosuppression too has been shown to be attributed with IL-33 [87]. Immune dysregulation in coronavirus infection is hypothesized to be caused by the IL-1 family member of cytokines [88]. IL-33 has also been shown to induce neutrophilic infiltration in several animal models and disease conditions, which may be interpreted as a pro-inflammatory effect [89].

IL-33 has dual nuclear and soluble cytokine forms. Nuclear IL-33 functions as a transcriptional regulator. In acute wound healing processes, IL-33 functions by attenuating inflammation by affecting the NF kappa B activity and enhancing wound healing [90]. On the other hand, IL-33 as a cytokine enhances immune reactions in decubitus ulcer models (unpublished). Both IL-33 and IL-1 alpha, when in the nucleus, bind to chromatin and are not released easily, thus, forming a reservoir for inflammatory signals. The regulation of nuclear or cytoplasmic localization of IL-33 is not clear but maybe dependent on its nuclear localization signal. Tsuda et al. [91] revealed that there are several different forms of splice variants of naturally occurring IL-33, of which expression is regulated by distinct promoters [92]. These splice variants should have distinct roles, which could regulate the pro- or anti-inflammatory properties of IL-33.

In the steady-state, keratinocytes should remain silent as a constitutively active state could result in excessive inflammation, which in turn can harm the overall human health. Cultured keratinocytes usually require higher concentrations of cytokines to provoke inflammatory signals; for example, keratinocytes need TNF in the range of several ng/ml to produce inflammatory cytokines, while dendritic cells or lymphocytes require only several pg/ml of the same cytokine to produce an inflammatory effect to the same or even a greater extent [93, 94]. Keratinocytes by differentiating to cornified cells become resistant to environmental stimuli, such as UVB; i.e., they usually respond sensitively to UVB in monolayer culture, but they become resistant to UVB stimulation when they differentiate in 3D-culture [95]. Some chemokines, such as MIP3 alpha/CCL20 are produced more in suprabasal cells than from basal cells [96], but production of IL-1 receptor antagonist is enhanced when keratinocytes are differentiated [97], which may result in attenuation of inflammatory response in differentiated keratinocytes. IL-33 and IL-1 alpha, more clearly expressed in suprabasal cells [98], when in the nucleus bind to chromatin not to be released easily, thus forming the reservoir for inflammatory signals.

7. Conclusion

Epidermal keratinocytes protect humans from the outer environment by forming an efficient mechanical, chemical, and antimicrobial barrier. Mutations in various molecules present in the keratinocyte can cause hereditary disorders. The keratinocyte structure is maintained by many structural molecules, including keratins, actin, microtubules, and associated proteins and adhesion molecules. The barrier function depends on these structural molecules, as well as other antimicrobial and immunological components, such as infiltrating or resident immune cells, such as lymphocytes, dendritic cells, and macrophages. At the same time, keratinocytes are resistant to stimulation in comparison to other cell types, such as lymphocytes and dendritic cells, as shown in some pieces of literature that they respond to the same stimuli with much fewer attitudes compared to immune cells. IL-33, an alarmin released during insults into the skin, works as an alarmin to provoke inflammation, but at the same time often attenuates inflammation by activating regulatory T cells and skewing Th2 mediated inflammation. This relative unresponsiveness and dual-faced character with pro- and anti-inflammatory properties would be the characteristics of keratinocytes, which cover the entire body by facing environmental stimuli all the time. Thus, the differentiation and structural characteristics of epidermal keratinocytes prevent the skin from hypersensitivity to environmental stimuli.

The mechanism of developing inflammatory conditions has been intensively investigated, but the mechanism by which the inflammation status returns to the steady-state, or how inflammatory status remains under control to prevent excessive inflammation in healthy humans has not been fully investigated.

A novel approach to maintaining healthy conditions would be unraveled when the mechanism of sequestrating inflammation and returning to normal steady-state condition is investigated. Epidermal keratinocytes are good candidates to investigate these mechanisms because they present both pro- and anti-inflammatory properties in vivo and in vitro.

Acknowledgements

I thank all the members of our department for participating in clinical and basic research on patients.

Conflict of interest

The authors declare no conflict of interest.

Author details

Mayumi Komine[1,2*], Jin Meijuan[1], Miho Kimura-Sashikawa[1], Razib MD. Hossain[1], Tuba M. Ansary[1], Tomoyuki Oshio[3], Jitlada Meephansan[4], Hidetoshi Tsuda[5], Shin-ichi Tominaga[2,6] and Mamitaro Ohtsuki[1]

1 Department of Dermatology, Jichi Medical University, Japan

2 Department of Biochemistry, Jichi Medical University, Japan

3 Ryukakusan, Co., Ltd., Japan

4 Division of Dermatology, Chulabhorn International College of Medicine, Thammasat University, Thailand

5 Division of Human Genetics, Center for Molecular Medicine, Jichi Medical University, Japan

6 Japan Association of Development of Community Medicine (JADECOM), Japan

*Address all correspondence to: mkomine12@jichi.ac.jp

IntechOpen

References

[1] Blumenberg M, Tomić-Canić M. Human epidermal keratinocyte: Keratinization processes. EXS. 1997;**78**:1-29. DOI: 10.1007/978-3-0348-9223-0_1

[2] Bragulla HH, Homberger DG. Structure and functions of keratin proteins in simple, stratified, keratinized and cornified epithelia. Journal of Anatomy. 2009;**214**(4):516-559. DOI: 10.1111/j.1469-7580.2009.01066.x

[3] Oji V, Tadini G, Akiyama M, Blanchet Bardon C, Bodemer C, Bourrat E, et al. Revised nomenclature and classification of inherited ichthyoses: Results of the first ichthyosis consensus conference in Sorèze 2009. Journal of the American Academy of Dermatology. 2010;**63**(4):607-641. DOI: 10.1016/j.jaad.2009.11.020

[4] Barker JN, Palmer CN, Zhao Y, Liao H, Hull PR, Lee SP, et al. Null mutations in the filaggrin gene (FLG) determine major susceptibility to early-onset atopic dermatitis that persists into adulthood. The Journal of Investigative Dermatology. 2007;**127**(3):564-567. DOI: 10.1038/sj.jid.5700587

[5] Henderson J, Northstone K, Lee SP, Liao H, Zhao Y, Pembrey M, et al. The burden of disease associated with filaggrin mutations: A population-based longitudinal birth cohort study. The Journal of Allergy and Clinical Immunology. 2008;**121**(4):872-7.e9. DOI: 10.1016/j.jaci.2008.01.026

[6] Kezic S, O'Regan GM, Lutter R, Jakasa I, Koster ES, Saunders S, et al. Filaggrin loss-of-function mutations are associated with enhanced expression of IL-1 cytokines in the stratum corneum of patients with atopic dermatitis and in a murine model of filaggrin deficiency. The Journal of Allergy and Clinical Immunology. 2012;**129**(4):1031-9.e1. DOI: 10.1016/j.jaci.2011.12.989

[7] WH ML. Filaggrin failure - from ichthyosis vulgaris to atopic eczema and beyond. The Journal of Dermatology. 2016;**175**(Suppl. 2):4-7. DOI: 10.1111/bjd.14997

[8] Eckert RL, Sturniolo MT, Broome AM, Ruse M, Rorke EA. Transglutaminases in epidermis. Progress in Experimental Tumor Research. 2005;**38**:115-124. DOI: 10.1159/000084236

[9] Richard G, Brown N, Rouan F, Van der Schroeff JG, Bijlsma E, Eichenfield LF, et al. Genetic heterogeneity in erythrokeratodermia variabilis: Novel mutations in the connexin gene GJB4 (Cx30.3) and genotype-phenotype correlations. The Journal of Investigative Dermatology. 2003;**120**(4):601-609. DOI: 10.1046/j.1523-1747.2003.12080.x

[10] Ishida-Yamamoto A, Kato H, Kiyama H, Armstrong DK, Munro CS, Eady RA, et al. Mutant loricrin is not crosslinked into the cornified cell envelope but is translocated into the nucleus in loricrin keratoderma. The Journal of Investigative Dermatology. 2000;**115**(6):1088-1094. DOI: 10.1046/j.1523-1747.2000.00163.x

[11] Ishida-Yamamoto A, Iizuka H. Structural organization of cornified cell envelopes and alterations in inherited skin disorders. Experimental Dermatology. 1998;**7**(1):1-10. DOI: 10.1111/j.1600-0625.1998.tb00295.x

[12] Epstein EH Jr, Leventhal ME. Steroid sulfatase of human leukocytes and epidermis and the diagnosis of recessive X-linked ichthyosis. The Journal of Clinical Investigation. 1981;**67**(5):1257-1262. DOI: 10.1172/jci110153

[13] Mazereeuw-Hautier J, Bitoun E, Chevrant-Breton J, Man SY, Bodemer C, Prins C, et al. Keratitis-ichthyosis-deafness syndrome: Disease expression and spectrum of connexin 26 (GJB2) mutations in 14 patients. The Journal of Dermatology. 2007;**156**(5):1015-1019. DOI: 10.1111/j.1365-2133.2007.07806.x

[14] Raghunath M, Tontsidou L, Oji V, Aufenvenne K, Schürmeyer-Horst F, Jayakumar A, et al. SPINK5 and Netherton syndrome: Novel mutations, demonstration of missing LEKTI, and differential expression of transglutaminases. The Journal of Investigative Dermatology. 2004;**123**(3): 474-483. DOI: 10.1111/j.0022-202X. 2004.23220.x

[15] Rizzo WB. Sjogren-Larsson syndrome: Molecular genetics and biochemical pathogenesis of fatty aldehyde dehydrogenase deficiency. Molecular Genetics and Metabolism. 2007;**90**(1):1-9. DOI: 10.1016/j. ymgme.2006.08.006

[16] Freedberg IM, Tomic-Canic M, Komine M, Blumenberg M. Keratins and the keratinocyte activation cycle. The Journal of Investigative Dermatology. 2001;**116**(5):633-640. DOI: 10.1046/j. 1523-1747.2001.01327.x

[17] Corden LD, McLean WH. Human keratin diseases: Hereditary fragility of specific epithelial tissues. Experimental Dermatology. 1996;**5**(6):297-307. DOI: 10.1111/j.1600-0625.1996. tb00133.x

[18] Smith FJ, Liao H, Cassidy AJ, Stewart A, Hamill KJ, Wood P, et al. The genetic basis of pachyonychia congenita. The Journal of Investigative Dermatology. Symposium Proceedings. 2005;**10**(1):21-30. DOI: 10.1111/j. 1087-0024.2005.10204.x

[19] Richard G, De Laurenzi V, Didona B, Bale SJ, Compton JG. Keratin 13 point mutation underlies the hereditary

mucosal epithelial disorder white sponge nevus. Nature Genetics. 1995;**11**(4):453-455. DOI: 10.1038/ ng1295-453

[20] Jang KH, Yoon HN, Lee J, Yi H, Park SY, Lee SY, et al. Liver disease-associated keratin 8 and 18 mutations modulate keratin acetylation and methylation. The FASEB Journal. 2019;**33**(8):9030-9043. DOI: 10.1096/ fj.201800263RR

[21] Zhou Q, Ji X, Chen L, Greenberg HB, Lu SC, Omary MB. Keratin mutation primes mouse liver to oxidative injury. Hepatology. 2005;**41**(3):517-525. DOI: 10.1002/ hep.20578

[22] Ku NO, Gish R, Wright TL, Omary MB. Keratin 8 mutations in patients with cryptogenic liver disease. The New England Journal of Medicine. 2001;**344**(21):1580-1587. DOI: 10.1056/ NEJM200105243442103

[23] Yancey KB. Adhesion molecules. II: Interactions of keratinocytes with epidermal basement membrane. The Journal of Investigative Dermatology. 1995;**104**(6):1008-1014. DOI: 10.1111/ 1523-1747.ep12606244

[24] Amagai M. Adhesion molecules. I: Keratinocyte-keratinocyte interactions; cadherins and pemphigus. The Journal of Investigative Dermatology. 1995;**104**(1):146-152. DOI: 10.1111/1523-1747.ep12613668

[25] Indra I, Hong S, Troyanovsky R, Kormos B, Troyanovsky S. The adherens junction: A mosaic of cadherin and nectin clusters bundled by actin filaments. The Journal of Investigative Dermatology. 2013;**133**(11):2546-2554. DOI: 10.1038/jid.2013.200

[26] Matsui T, Amagai M. Dissecting the formation, structure, and barrier function of the stratum corneum. International Immunology.

2015;**27**(6):269-280. DOI: 10.1093/intimm/dxv013

[27] Kitajima Y. 150(th) anniversary series: Desmosomes and autoimmune disease, perspective of dynamic desmosome remodeling and its impairments in pemphigus. Cell Communication & Adhesion. 2014;**21**(6):269-280. DOI: 10.3109/15419061.2014.943397

[28] Has C, Bauer JW, Bodemer C, Bolling MC, Bruckner-Tuderman L, Diem A, et al. Consensus reclassification of inherited epidermolysis bullosa and other disorders with skin fragility. The Journal of Dermatology. 2020;**183**(4): 614-627. DOI: 10.1111/bjd.18921

[29] Condrat I, He Y, Cosgarea R, Has C. Junctional epidermolysis bullosa: Allelic heterogeneity and mutation stratification for precision medicine. Front Med (Lausanne). 2019;**5**:363. DOI: 10.3389/fmed.2018.00363

[30] Chung HJ, Uitto J. Epidermolysis bullosa with pyloric atresia. Dermatologic Clinics. 2010;**28**(1):43-54. DOI: 10.1016/j.det.2009.10.005

[31] Iinuma S, Aikawa E, Tamai K, Fujita R, Kikuchi Y, Chino T, et al. Transplanted bone marrow-derived circulating PDGFRalpha+ cells restore type VII collagen in a recessive dystrophic epidermolysis bullosa mouse skin graft. Journal of Immunology. 2015;**194**(4):1996-2003. DOI: 10.4049/jimmunol.1400914

[32] Tamai K, Uitto J. Stem cell therapy for epidermolysis bullosa-does it work? The Journal of Investigative Dermatology. 2016;**136**(11):2119-2121. DOI: 10.1016/j.jid.2016.07.004

[33] McGrath JA, Bolling MC, Jonkman MF. Lethal acantholytic epidermolysis bullosa. Dermatologic Clinics. 2010;**28**(1):131-135. DOI: 10.1016/j.det.2009.10.015

[34] Horimukai K, Morita K, Narita M, Kondo M, Kitazawa H, Nozaki M, et al. Application of moisturizer to neonates prevents development of atopic dermatitis. The Journal of Allergy and Clinical Immunology. 2014;**134**(4):824-830.e6. DOI: 10.1016/j.jaci.2014.07.060

[35] Komine M. Analysis of the mechanism for the development of allergic skin inflammation and the application for its treatment: Keratinocytes in atopic dermatitis: Their pathogenic involvement. Journal of Pharmacological Sciences. 2009;**110**(3): 260-264. DOI: 10.1254/jphs.09r06fm

[36] Natsume O, Ohya Y. Recent advancement to prevent the development of allergy and allergic diseases and therapeutic strategy in the perspective of barrier dysfunction. Allergology International. 2018; **67**(1):24-31. DOI: 10.1016/j.alit.2017.11.003

[37] Mizuno O, Nomura T, Ohguchi Y, Suzuki S, Nomura Y, Hamade Y, et al. Loss-of-function mutations in the gene encoding filaggrin underlie a Japanese family with food-dependent exercise-induced anaphylaxis. Journal of the European Academy of Dermatology and Venereology. 2015;**29**(4):805-808. DOI: 10.1111/jdv.12441

[38] Yamamoto-Hanada K, Kobayashi T, Williams HC, Mikami M, Saito-Abe M, Morita K, et al. Early aggressive intervention for infantile atopic dermatitis to prevent development of food allergy: A multicenter, investigator-blinded, randomized, parallel group controlled trial (PACI study)-protocol for a randomized controlled trial. Clin Transl Allergy. 2018;**8**:47. DOI: 10.1186/s13601-018-0233-8

[39] Hovnanian A. Netherton syndrome: Skin inflammation and allergy by loss of protease inhibition. Cell and Tissue Research. 2013;**351**(2):289-300. DOI: 10.1007/s00441-013-1558-1

[40] Yamauchi A, Kubo A, Ono N, Shiohama A, Tsuruta D, Fukai K. Three cases of Nagashima-type palmoplantar keratosis were associated with atopic dermatitis: A diagnostic pitfall. The Journal of Dermatology. 2018;**45**(5): e112-e113. DOI: 10.1111/1346-8138. 14152

[41] Smith PK, Harper JI. Serine proteases, their inhibitors and allergy. Allergy. 2006;**61**(12):1441-1447. DOI: 10.1111/j.1398-9995.2006.01233.x

[42] Imokawa G, Abe A, Jin K, Higaki Y, Kawashima M, Hidano A. Decreased level of ceramides in stratum corneum of atopic dermatitis: An etiologic factor in atopic dry skin? The Journal of Investigative Dermatology. 1991;**96**(4): 523-526. DOI: 10.1111/1523-1747. ep12470233

[43] Teranishi Y, Kuwahara H, Ueda M, Takemura T, Kusumoto M, Nakamura K, et al. Sphingomyelin Deacylase, the enzyme involved in the pathogenesis of atopic dermatitis, is identical to the beta-subunit of acid ceramidase. International Journal of Molecular Sciences. 2020;**21**(22):8789. DOI: 10.3390/ijms21228789

[44] Cork MJ, Robinson DA, Vasilopoulos Y, Ferguson A, Moustafa M, MacGowan A, et al. New perspectives on epidermal barrier dysfunction in atopic dermatitis: Gene-environment interactions. The Journal of Allergy and Clinical Immunology. 2006;**118**(1):3-21; quiz 22-3. DOI: 10.1016/j.jaci.2006.04.042

[45] Takahashi T, Gallo RL. The critical and multifunctional roles of antimicrobial peptides in dermatology. Dermatologic Clinics. 2017;**35**(1):39-50. DOI: 10.1016/j.det.2016.07.006

[46] Ong PY, Ohtake T, Brandt C, Strickland I, Boguniewicz M, Ganz T, et al. Endogenous antimicrobial peptides and skin infections in atopic dermatitis. The New England Journal of Medicine. 2002;**347**(15):1151-1160. DOI: 10.1056/NEJMoa021481

[47] Smits JPH, Ederveen THA, Rikken G, van den Brink NJM, van Vlijmen-Willems IMJJ, Boekhorst J, et al. Targeting the cutaneous microbiota in atopic dermatitis by coal tar via AHR-dependent induction of antimicrobial peptides. The Journal of Investigative Dermatology. 2020;**140**(2):415-424.e10. DOI: 10.1016/j.jid.2019.06.142

[48] Rangel SM, Paller AS. Bacterial colonization, overgrowth, and superinfection in atopic dermatitis. Clinics in Dermatology. 2018;**36**(5):641-647. DOI: 10.1016/j.clindermatol. 2018.05.005

[49] Morizane S, Gallo RL. Antimicrobial peptides in the pathogenesis of psoriasis. The Journal of Dermatology. 2012;**39**(3):225-230. DOI: 10.1111/j. 1346-8138.2011.01483.x

[50] Chiricozzi A, Nograles KE, Johnson-Huang LM, Fuentes-Duculan J, Cardinale I, Bonifacio KM, et al. IL-17 induces an expanded range of downstream genes in reconstituted human epidermis model. PLoS One. 2014;**9**(2):e90284. DOI: 10.1371/journal. pone.0090284

[51] Lande R, Gregorio J, Facchinetti V, Chatterjee B, Wang YH, Homey B, et al. Plasmacytoid dendritic cells sense self-DNA coupled with antimicrobial peptide. Nature. 2007;**449**(7162):564-569. DOI: 10.1038/nature06116

[52] Ganguly D, Chamilos G, Lande R, Gregorio J, Meller S, Facchinetti V, et al. Self-RNA-antimicrobial peptide complexes activate human dendritic cells via TLR7 and TLR8. The Journal of Experimental Medicine. 2009;**206**(9): 1983-1994. DOI: 10.1084/jem.20090480

[53] Gordon K, Papp K, Poulin Y, Gu Y, Rozzo S, Sasso EH. Long-term efficacy

and safety of adalimumab in patients with moderate to severe psoriasis treated continuously over 3 years: Results from an open-label extension study for patients from REVEAL. Journal of the American Academy of Dermatology. 2012;**66**(2):241-251. DOI: 10.1016/j.jaad.2010.12.005

[54] Chaudhari U, Romano P, Mulcahy LD, Dooley LT, Baker DG, Gottlieb AB. Efficacy and safety of infliximab monotherapy for plaque-type psoriasis: A randomised trial. Lancet. 2001;**357**(9271):1842-1847. DOI: 10.1016/s0140-6736(00)04954-0

[55] Reich K, Ortonne JP, Gottlieb AB, Terpstra IJ, Coteur G, Tasset C, et al. Successful treatment of moderate to severe plaque psoriasis with the PEGylated Fab' certolizumab pegol: Results of a phase II randomized, placebo-controlled trial with a re-treatment extension. The British Journal of Dermatology. 2012;**167**(1): 180-190. DOI: 10.1111/j.1365-2133. 2012.10941.x

[56] Ohtsuki M, Morita A, Abe M, Takahashi H, Seko N, Karpov A, et al. Secukinumab efficacy and safety in Japanese patients with moderate-to-severe plaque psoriasis: Subanalysis from ERASURE, a randomized, placebo-controlled, phase 3 study. The Journal of Dermatology. 2014;**41**(12): 1039-1046. DOI: 10.1111/1346-8138. 12668

[57] Langley RG, Elewski BE, Lebwohl M, Reich K, Griffiths CE, Papp K, et al. Secukinumab in plaque psoriasis--results of two phase 3 trials. The New England Journal of Medicine. 2014;**371**(4):32638. DOI: 10.1056/ NEJMoa1314258

[58] Gordon KB, Blauvelt A, Papp KA, Langley RG, Luger T, Ohtsuki M, et al. Phase 3 trials of Ixekizumab in moderate-to-severe plaque psoriasis. The New England Journal of Medicine.

2016;**375**(4):345-356. DOI: 10.1056/ NEJMoa1512711

[59] Papp K, Leonardi C, Menter A, Thompson EH, Milmont CE, Kricorian G, et al. Safety and efficacy of brodalumab for psoriasis after 120 weeks of treatment. Journal of the American Academy of Dermatology. 2014;**71**(6):1183-1190.e3. DOI: 10.1016/j. jaad.2014.08.039

[60] Blauvelt A, Papp KA, Merola JF, Gottlieb AB, Cross N, Madden C, et al. Bimekizumab for patients with moderate to severe plaque psoriasis: 60-week results from BE ABLE 2, a randomized, double-blinded, placebo-controlled, phase 2b extension study. Journal of the American Academy of Dermatology. 2020;**83**(5):1367-1374. DOI: 10.1016/j.jaad.2020.05.105

[61] Leonardi CL, Kimball AB, Papp KA, Yeilding N, Guzzo C, Wang Y, et al. Efficacy and safety of ustekinumab, a human interleukin-12/23 monoclonal antibody, in patients with psoriasis: 76-week results from a randomised, double-blind, placebo-controlled trial (PHOENIX 1). Lancet. 2008;**371**(9625): 1665-1674. DOI: 10.1016/S0140-6736 (08)60725-4

[62] Papp KA, Langley RG, Lebwohl M, Krueger GG, Szapary P, Yeilding N, et al. Efficacy and safety of ustekinumab, a human interleukin-12/23 monoclonal antibody, in patients with psoriasis: 52-week results from a randomised, double-blind, placebo-controlled trial (PHOENIX 2). Lancet. 2008;**371**(9625): 1675-1684. DOI: 10.1016/S0140-6736 (08)60726-6

[63] Blauvelt A, Papp KA, Griffiths CE, Randazzo B, Wasfi Y, Shen YK, et al. Efficacy and safety of guselkumab, an anti-interleukin-23 monoclonal antibody, compared with adalimumab for the continuous treatment of patients with moderate to severe psoriasis: Results from the phase III,

double-blinded, placebo- and active comparator-controlled VOYAGE 1 trial. Journal of the American Academy of Dermatology. 2017;**76**(3):405-417. DOI: 10.1016/j.jaad.2016.11.041

[64] Gordon KB, Strober B, Lebwohl M, Augustin M, Blauvelt A, Poulin Y, et al. Efficacy and safety of risankizumab in moderate-to-severe plaque psoriasis (UltIMMa-1 and UltIMMa-2): Results from two double-blind, randomised, placebo-controlled and ustekinumab-controlled phase 3 trials. Lancet. 2018;**392**(10148):650-661. DOI: 10.1016/ S0140-6736(18)31713-6

[65] Ohtsuki M, Fujita H, Watanabe M, Suzaki K, Flack M, Huang X, et al. Efficacy and safety of risankizumab in Japanese patients with moderate to severe plaque psoriasis: Results from the SustaIMM phase 2/3 trial. The Journal of Dermatology. 2019;**46**(8):686-694. DOI: 10.1111/1346-8138.14941

[66] Reich K, Papp KA, Blauvelt A, Tyring SK, Sinclair R, Thaçi D, et al. Tildrakizumab versus placebo or etanercept for chronic plaque psoriasis (reSURFACE 1 and reSURFACE 2): Results from two randomised controlled, phase 3 trials. Lancet. 2017;**390**(10091):276-288

[67] Reich K, Warren RB, Iversen L, Puig L, Pau-Charles I, Igarashi A, et al. Long-term efficacy and safety of tildrakizumab for moderate-to-severe psoriasis: Pooled analyses of two randomized phase III clinical trials (reSURFACE 1 and reSURFACE 2) through 148 weeks. The British Journal of Dermatology. 2020;**182**(3):605-617. DOI: 10.1111/bjd.18232

[68] Imafuku S, Nakagawa H, Igarashi A, Morita A, Okubo Y, Sano S, et al. Long-term efficacy and safety of tildrakizumab in Japanese patients with moderate to severe plaque psoriasis: Results from a 5-year extension of a phase 3 study (reSURFACE 1). The

Journal of Dermatology. 2021;**48**(6):844-852. DOI: 10.1111/ 1346-8138.15763

[69] Igarashi A, Nakagawa H, Morita A, Okubo Y, Sano S, Imafuku S, et al. Long-term efficacy and safety of tildrakizumab in Japanese patients with moderate to severe plaque psoriasis: Results from a 5-year extension of a phase 3 study (reSURFACE 1). The Journal of Dermatology. 2021;**48**(6):853-863. DOI: 10.1111/1346-8138.15789

[70] Tanaka Y, Luo Y, O'Shea JJ, Nakayamada S. Janus kinase-targeting therapies in rheumatology: A mechanisms-based approach. Nature Reviews Rheumatology. 2022:1-13. DOI: 10.1038/s41584-021-00726-8

[71] Papp KA, Menter MA, Abe M, Elewski B, Feldman SR, Gottlieb AB, et al. Tofacitinib, an oral Janus kinase inhibitor, for the treatment of chronic plaque psoriasis: Results from two randomized, placebo-controlled, phase III trials. The British Journal of Dermatology. 2015;**173**(4):949-961. DOI: 10.1111/bjd.14018

[72] Asahina A, Etoh T, Igarashi A, Imafuku S, Saeki H, Shibasaki Y, et al. Oral tofacitinib efficacy, safety and tolerability in Japanese patients with moderate to severe plaque psoriasis and psoriatic arthritis: A randomized, double-blind, phase 3 study. The Journal of Dermatology. 2016;**43**(8):869-880. DOI: 10.1111/1346-8138.13258

[73] Papp KA, Menter MA, Raman M, Disch D, Schlichting DE, Gaich C, et al. A randomized phase 2b trial of baricitinib, an oral Janus kinase (JAK) 1/JAK2 inhibitor, in patients with moderate-to-severe psoriasis. The British Journal of Dermatology. 2016;**174**(6):1266-1276. DOI: 10.1111/ bjd.14403

[74] Mease PJ, Lertratanakul A, Papp KA, van den Bosch FE, Tsuji S,

Dokoupilova E, et al. Upadacitinib in patients with psoriatic arthritis and inadequate response to biologics: 56-week data from the randomized controlled phase 3 SELECT-PsA 2 study. Rheumatol Ther. 2021;**8**(2):903-919. DOI: 10.1007/s40744-021-00305-z

[75] Schmieder GJ, Draelos ZD, Pariser DM, Banfield C, Cox L, Hodge M, et al. Efficacy and safety of the Janus kinase 1 inhibitor PF-04965842 in patients with moderate-to-severe psoriasis: Phase II, randomized, double-blind, placebo-controlled study. The British Journal of Dermatology. 2018;**179**(1):54-62. DOI: 10.1111/bjd.16004

[76] Nogueira M, Puig L, Torres T. JAK inhibitors for treatment of psoriasis: Focus on selective TYK2 inhibitors. Drugs. 2020;**80**(4):341-352. DOI: 10.1007/s40265-020-01261-8

[77] Malik K, He H, Huynh TN, Tran G, Mueller K, Doytcheva K, et al. Ichthyosis molecular fingerprinting shows profound T_H17 skewing and a unique barrier genomic signature. The Journal of Allergy and Clinical Immunology. 2019;**143**(2):604-618. DOI: 10.1016/j.jaci.2018.03.021

[78] Sheetz K, Lynch PJ. Ichthyosis and dermatophyte fungal infection. Journal of the American Academy of Dermatology. 1991;**24**(2 Pt 1):321. DOI: 10.1016/s0190-9622(08)80637-8

[79] Schøsler L, Andersen LK, Arendrup MC, Sommerlund M. Recurrent terbinafine resistant Trichophyton rubrum infection in a child with congenital ichthyosis. Pediatric Dermatology. 2018;**35**(2):259-260. DOI: 10.1111/pde.13411

[80] Cayrol C, Girard JP. Interleukin-33 (IL-33): A nuclear cytokine from the IL-1 family. Immunological Reviews. 2018;**281**(1):154-168. DOI: 10.1111/imr.12619

[81] Groves RW, Sherman L, Mizutani H, Dower SK, Kupper TS. Detection of interleukin-1 receptors in the human epidermis induction of the type II receptor after organ culture and psoriasis. The American Journal of Pathology. 1994;**145**(5):1048-1056

[82] Ross R, Grimmel J, Goedicke S, Möbus AM, Bulau AM, Bufler P, et al. Analysis of the nuclear localization of interleukin-1 family cytokines by flow cytometry. Journal of Immunological Methods. 2013;**387**(1-2):219-227. DOI: 10.1016/j.jim.2012.10.017

[83] Hayakawa H, Hayakawa M, Kume A, Tominaga S. Soluble ST2 blocks interleukin-33 signaling in allergic airway inflammation. The Journal of Biological Chemistry. 2007;**282**(36):26369-26380. DOI: 10.1074/jbc.M704916200

[84] Matta BM, Reichenbach DK, Zhang X, Mathews L, Koehn BH, Dwyer GK, et al. Peri-alloHCT IL-33 administration expands the recipient T-regulatory cells that protect mice against acute GVHD. Blood. 2016;**128**(3):427-439. DOI: 10.1182/blood-2015-12-684142

[85] Jiang HR, Milovanović M, Allan D, Niedbala W, Besnard AG, Fukada SY, et al. IL-33 attenuates EAE by suppressing IL-17 and IFN-γ production and inducing alternatively activated macrophages. European Journal of Immunology. 2012;**42**(7):1804-1814. DOI: 10.1002/eji.201141947

[86] Matta BM, Lott JM, Mathews LR, Liu Q, Rosborough BR, Blazar BR, et al. IL-33 is an unconventional Alarmin that stimulates IL-2 secretion by dendritic cells to selectively expand IL-33R/ST2+ regulatory T cells. Journal of Immunology. 2014;**193**(8):4010-4020. DOI: 10.4049/jimmunol.1400481

[87] Byrne SN, Beaugie C, O'Sullivan C, Leighton S, Halliday GM. The

immune-modulating cytokine and endogenous Alarmin interleukin-33 is upregulated in skin exposed to inflammatory UVB radiation. The American Journal of Pathology. 2011;**179**(1):211-222. DOI: 10.1016/j.ajpath.2011.03.010

[88] Kritas SK, Ronconi G, Caraffa A, Gallenga CE, Ross R, Conti P. Mast cells contribute to coronavirus-induced inflammation: New anti-inflammatory strategy. Journal of Biological Regulators and Homeostatic Agents. 2020;**34**(1):9-14. DOI: 10.23812/20-Editorial-Kritas

[89] Alves-Filho JC, Sônego F, Souto FO, Freitas A, Verri WA Jr, Auxiliadora-Martins M, et al. Interleukin-33 attenuates sepsis by enhancing neutrophil influx to the site of infection. Nature Medicine. 2010;**16**(6):708-712. DOI: 10.1038/nm.2156

[90] Oshio T, Komine M, Tsuda H, Tominaga SI, Saito H, Nakae S, et al. Nuclear expression of IL-33 in epidermal keratinocytes promotes wound healing in mice. Journal of Dermatological Science. 2017;**85**(2):106-114. DOI: 10.1016/j.jdermsci.2016.10.008

[91] Tsuda H, Komine M, Karakawa M, Etoh T, Tominaga S, Ohtsuki M. Novel splice variants of IL-33: Differential expression in normal and transformed cells. The Journal of Investigative Dermatology. 2012;**132**(11):2661-2664. DOI: 10.1038/jid.2012.180

[92] Tsuda H, Komine M, Tominaga SI, Ohtsuki M. Identification of the promoter region of human IL-33 responsive to induction by IFNγ. Journal of Dermatological Science. 2017;**85**(2):137-140. DOI: 10.1016/j.jdermsci.2016.11.002

[93] Chung JH, Youn SH, Koh WS, Eun HC, Cho KH, Park KC, et al. Ultraviolet B irradiation-enhanced interleukin (IL)-6 production and mRNA expression are mediated by IL-1 alpha in cultured human keratinocytes. The Journal of Investigative Dermatology. 1996;**106**(4):715-720. DOI: 10.1111/1523-1747.ep12345608

[94] Tosato G, Jones KD. Interleukin-1 induces interleukin-6 production in peripheral blood monocytes. Blood. 1990;**75**(6):1305-1310

[95] Corsini E, Sangha N, Feldman SR. Epidermal stratification reduces the effects of UVB (but not UVA) on keratinocyte cytokine production and cytotoxicity. Photodermatology, Photoimmunology & Photomedicine. 1997;**13**(4):147-152. DOI: 10.1111/j.1600-0781.1997.tb00219.x

[96] Tohyama M, Shirakara Y, Yamasaki K, Sayama K, Hashimoto K. Differentiated keratinocytes are responsible for TNF-alpha regulated production of macrophage inflammatory protein 3alpha/CCL20, a potent chemokine for Langerhans cells. Journal of Dermatological Science. 2001;**27**(2):130-139. DOI: 10.1016/s0923-1811(01)00127-x

[97] Bigler CF, Norris DA, Weston WL, Arend WP. Interleukin-1 receptor antagonist production by human keratinocytes. The Journal of Investigative Dermatology. 1992;**98**(1):38-44. DOI: 10.1111/1523-1747.ep12494196

[98] Meephansan J, Komine M, Tsuda H, Karakawa M, Tominaga S, Ohtsuki M. Expression of IL-33 in the epidermis: The mechanism of induction by IL-17. Journal of Dermatological Science. 2013;**71**(2):107-114. DOI: 10.1016/j.jdermsci.2013.04.014 Epub 2013 Apr 19

www.ingramcontent.com/pod-product-compliance
Lightning Source LLC
Chambersburg PA
CBHW081229190326
41458CB00016B/5733